ISBN 978-0-331-66664-9
PIBN 11069104

PROPOSED
TARIFF REVISION LAW
OF 1909 FOR THE PHILIPPINE
ISLANDS   /53

MESSAGE FROM THE PRESIDENT OF THE
UNITED STATES, TRANSMITTING A COM-
MUNICATION FROM THE SECRETARY OF
WAR IN REFERENCE TO THE PROPOSED
TARIFF REVISION LAW OF 1909 FOR THE
PHILIPPINE ISLANDS

APRIL 15, 1909.—Read; referred to the Committee on Ways and Means
and ordered to be printed

WASHINGTON
GOVERNMENT PRINTING OFFICE
1909

PROPOSED TARIFF REVISION LAW OF 1909 FOR THE PHILIPPINE ISLANDS.

To the Senate and House of Representatives:

I transmit herewith a communication from the Secretary of War, inclosing one from the Chief of the Bureau of Insular Affairs, in which is transmitted a proposed tariff revision law for the Philippine Islands.

This measure revises the present Philippine tariff, simplifies it, and makes it conform as nearly as possible to the regulations of the customs laws of the United States, especially with respect to packing and packages. The present Philippine regulations have been cumbersome and difficult for American merchants and exporters to comply with. Its purpose is to meet the new conditions that will arise under the section of the pending United States tariff bill which provides, with certain limitations, for free trade between the United States and the islands. It is drawn with a view to preserving to the islands as much customs revenue as possible, and to protect in a reasonable measure those industries which now exist in the islands.

The bill now transmitted has been drawn by a board of tariff experts, of which the insular collector of customs, Col. George R. Colton, was the president. The board held a great many open meetings in Manila, and conferred fully with representatives of all business interests in the Philippine Islands. It is of great importance to the welfare of the islands that the bill should be passed at the same time with the pending Payne bill, with special reference to the provisions of which it was prepared.

I respectfully recommend that this bill be enacted at the present session of Congress as one incidental to and required by the passage of the Payne bill.

WM. H. TAFT.

THE WHITE HOUSE, *April 14, 1909.*

WAR DEPARTMENT,
Washington, April 3, 1909.

MY DEAR MR. PRESIDENT: I inclose herewith a proposed tariff-revision act for the Philippine Islands and request that the same be transmitted to Congress in such manner as may be deemed advisable with a view to its passage at the present session.

I also inclose herewith a letter to me from General Edwards, the Chief of the Bureau of Insular Affairs, which explains the matter in detail.

3

I have not had time to examine the bill in detail and have not sufficient acquaintance with the subject to say whether or not it is what it should be; but General Edwards, who is familiar with the matter, recommends it, and I have no doubt that with your own familiarity with the subject you will be able to dispose of it.

Sincerely, yours,

J. M. DICKINSON.

The PRESIDENT.

WAR DEPARTMENT,
BUREAU OF INSULAR AFFAIRS,
Washington, April 1, 1909.

SIR: I have the honor to transmit herewith a proposed tariff revision act for the Philippine Islands, and request that the same be transmitted to Congress in such manner as may be deemed advisable, with a view to its passage at the present session.

This measure is a complete revision of the present Philippine tariff, drafted with a view to simplifying the law and making it conform as nearly as possible to the customs laws of the United States, especially with respect to packing and packages, in which particular the present Philippine regulations are cumbersome and difficult to comply with.

The purpose of the bill is also to meet the conditions which will arise under the free-trade provisions of the Payne bill, to save as much customs revenue as possible to the islands under those conditions, and at the same time to protect in reasonable measure the paying industries now operating in the islands.

It will be understood that the result of the free admission of American goods into the Philippine Islands must revolutionize business in the Philippines, and unless the adoption of that policy is accompanied by a revision of the present Philippine tariff it will be disastrous to some important industries in the islands and also result in such serious loss to the customs revenue as to embarrass the Philippine government.

The bill now presented has been in course of preparation, by special direction of the Secretary of War, since March 1, 1908, and was finally drawn by a board of tariff experts of which the insular collector of customs, Col. George R. Colton, was the president, after numerous open meetings in Manila and full conference with representatives of all interests in the Philippine Islands to be affected thereby.

It has had publicity in the United States, and it is believed that contending interests have been reconciled, and as far as schedules are concerned it should meet with no opposition. For instance, the schedules relating to the introduction of tobacco, Sumatra leaf, and sugar are made identical with the pending Payne bill, and therefore have removed the apprehension that these goods can be imported into the Philippine Islands at a less tariff rate and thence into the United States free as the growth and product of the Philippine Islands.

The proposed revision has the approval of the governor-general and the government he represents in the Philippine Islands.

It is therefore respectfully suggested that this bill be presented as soon as practicable and the importance of its passage at this session of Congress, for the reasons stated, be impressed upon that body.

Very respectfully.

C. R. EDWARDS.
Brigadier-General. U. S. Army,
Chief of Bureau.

The SECRETARY OF WAR.

PROPOSED TARIFF REVISION LAW OF 1909, PHILIPPINE ISLANDS.

AN ACT To revise and amend the tariff laws of the Philippine Islands, and for other purposes.

Be it enacted by the Senate and House of Representatives of the United States of America in Congress assembled. That the provisions of the act of Congress of the United States of America of March 3, 1905, entitled "An act to revise and amend the tariff laws of the Philippine Islands, and for other purposes," relating to customs duties on imports into and exports from the said Philippine Islands, and tonnage dues and wharf charges within the jurisdiction of the same, together with all acts, orders, proclamations, or decrees, or parts of acts, orders, proclamations, or decrees, amendatory thereof, or treating of any of the subjects therein treated of, or which are or may be mentioned, set out, or treated of in this act. either made, passed, issued, or promulgated prior to the date this act becomes effective, be, and hereby are, amended to read as follows, subject, however, to the provisions of section 5 of the act of Congress approved ———, 1909, known as the Payne bill:

SEC. 2. That duties shall be collected on all articles, goods, wares, or merchandise imported into the Philippine Islands at the rates hereinafter provided, except when same are expressly exempted from duty by this act. Any articles, goods, wares, or merchandise from abroad, except as hereinafter provided, entering the jurisdiction of the Philippine Islands, in any manner whatsoever. either with intent to unlade therein, or which, after such entering, become incorporated into the general mass of property within said islands, or are consumed therein, shall be deemed to have been imported within the meaning of this act.

SEC. 3. That articles, goods, wares. or merchandise in transit to the Philippine Islands at the time this act goes into effect. and arriving therein not later than sixty days from such effective date, shall be entered under the provisions of law existing at the time of shipment from the place of original consignment to the Philippine Islands.

SEC. 4. That duties shall be collected on such articles, goods, wares, or merchandise exported from the Philippine Islands and at such rates as are hereinafter specifically prescribed in this act, but that exportation from the said islands of all other articles, goods, wares, or merchandise shall be free.

SEC. 5. That the following rules shall be observed in the construction and enforcement of the various provisions of this act:

GENERAL RULES.

TREATMENT OF TEXTILES.

RULE 1. NUMBER OF THREADS AND ASCERTAINMENT THEREOF.—By the number of threads in a textile shall, unless otherwise stipulated, be meant the total number of all threads contained in the warp and weft thereof in a square of six millimeters. Warp is the total number of threads which lie longitudinally in a textile, whether they form the foundation thereof or have been added thereto. Weft shall be considered the total number of threads which cross the warp, whether from selvage to selvage or not. To determine the number of threads in a textile, and the proportion thereof subject to the highest rate of duty, a "thread counter" shall be used.

Should a textile be more closely woven in some parts than in others, the number of threads in the most closely woven part and in the most loosely woven part shall be ascertained, and the average number of threads resulting shall serve as the basis for levying duty.

Threads shall be counted on the finished side of the textile, if the nature thereof permits; otherwise, on the reverse side. If necessary, to ascertain the number of threads, the nap shall be removed or a sufficient part of the textile unraveled.

Should this be impossible without damaging a made-up article, as, for instance, ready-made clothing, the textile shall be subject to the highest rate of duty applicable in the group to which it belongs, and if the textile be mixed, it shall be dutiable at the rate applicable to the most highly taxed component material in the exterior of the article.

RULE 2. SURTAXES: (a) HOW COMPUTED.—The surtaxes applicable on account of broché, metal threads, embroidery, trimming, or making-up, shall be computed on the primary duties leviable on the textile, including therewith the increase of such duties in case, and on account, of admixture.

(b) ON GOODS DUTIABLE AD VALOREM. Articles of any character, dutiable at an ad valorem rate, shall not be subject to any of the surtaxes provided herein, unless the application of such surtaxes to said ad valorem rate is specifically provided for in this act.

RULE 3. ADMIXTURES OF TWO MATERIALS.—Textiles composed of two materials shall be dutiable as follows:

(a) Cotton textiles containing threads of other vegetable fibers, and in which the total number of such threads, counted in the warp and weft, does not exceed one-fifth of the total number of threads composing the textile, shall be dutiable under the corresponding paragraphs of Class V, with a surtax of fifteen per centum.

When the number of threads of other vegetable fibers exceeds one-fifth of the total, the textile shall be dutiable under the corresponding paragraph of Class VI.

Cotton textiles containing threads of wool, hair, or wastes of these materials, and in which the total number of such threads, counted in the warp and weft, does not exceed one-fifth of the total number of threads composing the textile, shall be dutiable under the corresponding paragraphs of Class V, with a surtax of thirty-five per centum.

When the number of threads of wool, hair, or their wastes exceeds one-fifth of the total, the textile shall be dutiable under the corresponding paragraph of Class VII.

Cotton textiles containing threads of silk, and in which the total number of such threads, counted in the warp and weft, does not exceed one-fifth of the total number of threads composing the textile, shall be dutiable under the corresponding paragraphs of Class V, with a surtax of seventy per centum.

When the number of threads of silk exceeds one-fifth of the total, the textile shall be dutiable under the corresponding paragraph of Class VIII.

The provisions of this rule shall not apply to pile fabrics, knitted or netted stuffs, tulles, laces, or blondes (rule 6), or to ribbons, galloons, braids, tape, or trimmings (rule 7).

(b) Textiles of vegetable fibers (except cotton), containing threads of wool, hair, or their wastes, and in which the number of such threads, counted in the warp and weft, does not exceed one-fifth of the total number of threads composing the textile, shall be dutiable under the corresponding paragraphs of Class VI, with a surtax of forty per centum.

When the number of threads of wool, hair, or their wastes exceeds one-fifth of the total, the textile shall be dutiable under the corresponding paragraph of Class VII.

Textiles of vegetable fibers (except cotton), containing threads of silk, and in which the number of such threads, counted in the warp and weft, does not exceed one-fifth of the total number of threads composing the textile, shall be dutiable under the corresponding paragraphs of Class VI, with a surtax of sixty per centum.

When the number of silk threads exceeds one-fifth of the total, the textile shall be dutiable under the corresponding paragraphs of Class VIII.

The provisions of this rule shall not apply to pile fabrics, knitted or netted stuffs, tulles, laces, or blondes (rule 6), or to ribbons, galloons, braids, tape, or trimmings (rule 7).

(c) Textiles of wool or hair, containing threads of silk, and in which the number of such threads exceeds one-fifth of the total number of threads composing the textile, shall be dutiable under the corresponding paragraphs of Class VIII.

RULE 4. ADMIXTURES OF MORE THAN TWO MATERIALS.— Textiles composed of more than two materials shall be dutiable as follows:

(a) Textiles of an admixture of wool and cotton, or of wool and other vegetable fibers, containing threads of silk, but in which the number of silk threads, counted in the warp and weft, does not exceed one-fifth of the total number of threads composing the textile, shall be dutiable under the corresponding paragraphs of Class VII.

When the number of silk threads exceeds one-fifth of the total, the textile shall be dutiable under the corresponding paragraph of Class VIII.

(b) Textiles of an admixture of cotton and other vegetable fibers, together with threads of silk, but in which the number of silk threads, counted in the warp and weft, does not exceed one-fifth of the total number of threads composing the textile, shall be dutiable under the corresponding paragraphs of Class VI, and in addition, shall be liable to a surtax of seventy per centum for the threads of silk.

When the number of silk threads exceeds one-fifth of the total, the textile shall be dutiable under the corresponding paragraph of Class VIII.

(c) Textiles of an admixture of wool, cotton, and other vegetable fibers, containing no silk threads, and in which the number of threads of wool, counted in the warp and weft, does not exceed one-fifth of the total number of threads composing the textile, shall be dutiable under the corresponding paragraphs of Class VI, and in addition, shall be liable to a surtax of forty per centum for the threads of wool.

When the number of threads of wool exceeds one-fifth of the total, the textile shall be dutiable under the corresponding paragraph of Class VII.

RULE 5. SILK TEXTILES. All textiles containing silk threads, the number of which, counted in the warp and weft, exceeds one-fifth of the total number of threads composing the textile, shall be deemed textiles of silk.

EXCEPTIONS.

RULE 6. PILE FABRICS, AND KNITTED AND NETTED STUFFS.— Plushes, velvets, velveteens, and all pile fabrics: all kinds of knitted or netted stuffs; tulles, laces, and blondes; containing an admixture of textile materials, shall be dutiable at the rate applicable to the most highly taxed component material, whatever be the proportion of such material in the article.

RULE 7. RIBBONS, GALLOONS, BRAIDS, TAPE, AND TRIMMINGS.— Ribbons, galloons, braids, tape, and trimmings, containing an admixture of textile materials, shall be dutiable at the rate applicable to the most highly taxed component material, whatever be the proportion of such material in the article. When any of these articles contain metal threads in any proportion they shall be dutiable under the corresponding paragraphs of Class VIII.

RULE 8. BROCHÉS. Brochés dutiable under Class V, with silk, shall be liable to the duties leviable thereon with a surtax of fifteen per centum.

Brochés, dutiable under Class VI, with silk, shall be liable to the duties leviable thereon with a surtax of thirty per centum.

Brochés are textiles with ornamental figures formed by means of a shuttle at time of weaving, and in such manner that the threads forming the figure occupy only the space thereof.

RULE 9. EMBROIDERY AND TRIMMINGS.—Textiles, embroidered by hand or machine after weaving, or with application of trimmings, shall be liable to the duties leviable thereon with a surtax of thirty per centum.

If the embroidery contains threads of purl or common metals or of silver or spangles of any material other than gold, the surtax shall be sixty per centum of the duties applicable to the textile.

When the threads, purl, or spangles are of gold, the surtax shall be one hundred per centum.

Embroidery is distinguished from patterns woven in the textile by the latter being destroyed by unraveling the weft of the textile, while embroidery is independent of the warp and weft and can not be unraveled.

RULE 10. METALLIC THREADS.—Textiles composed exclusively of metallic threads shall be dutiable under Class VIII.

Textiles or articles (except those provided for in rules 7 and 9 hereof), dutiable under Classes V and VI, containing threads or purl of common metals or of silver, shall be liable to a surtax of fifty per centum of the duties leviable thereon.

If the threads or purl are of gold the surtax shall be one hundred per centum.

RULE 11. MADE-UP ARTICLES.—Textiles, dutiable under Classes V and VI, entirely or partially made up into common sacks (except gunny sacks), shall be liable to the duties applicable thereto with a surtax of fifteen per centum.

Shawls, including those called "mantones" and "pañolones," traveling rugs, sarongs, patadeones, counterpanes, sheets, towels, table cloths and napkins, veils, fichus and handkerchiefs, shall, for the making-up, be liable to a surtax of thirty per centum of the duties leviable thereon. Any of these articles, imported in the piece, uncut, shall not be considered as made-up, except in those cases where the line of separation between them is indicated by unwoven spaces.

Other articles, including wearing apparel, not otherwise provided for, cut, basted, partially finished, or finished, shall be treated in accordance with rule 1, and shall be dutiable at the rate applicable to the most highly taxed component material in the exterior thereof, with a surtax of fifty per centum.

ARTICLES NOT ENUMERATED AND THOSE COMPOSED OF SEVERAL MATERIALS.

RULE 12. On any article, not enumerated in this act, manufactured of two or more materials, duty shall be assessed at the rate at which the same would be dutiable if composed wholly of the component material thereof of chief value; and the words "component material of chief value," wherever used in this act, shall be held to mean that component material which shall exceed in value any other single component material of the article; and the value of each component material shall be determined by the ascertained value of such material in its condition as found in the article.

(b) If two or more rates of duty shall be applicable to any article, it shall pay duty at the highest of such rates.

(c) No customs officer shall give an advance opinion as to the classification for duty of any article intended to be imported: *Provided,* That when an article intended to be imported is not specifically mentioned in this act, the interested party or the importer may deposit with the insular collector of customs a sample thereof and request him to indicate the paragraph under which the article is or shall be dutiable, and the insular collector of customs shall comply with such request. In such case classification of the article in question, upon the particular importation involved, shall be made according to the paragraph so indicated.

(d) Salvage from vessels built in foreign countries and wrecked or abandoned in Philippine waters or elsewhere, not otherwise provided for, shall be dutiable according to the corresponding paragraphs of this act.

RECEPTACLES, PACKAGES, AND PACKING.

RULE 13. (a) Whenever imported merchandise is subject to an ad valorem rate of duty, the duty shall be assessed upon the actual market value or wholesale price of such merchandise, as bought and sold in usual wholesale quantities, at the time of exportation to the Philippine Islands, in the principal markets of the country from whence imported, and in the condition in which such merchandise is there bought and sold for exportation to the Philippine Islands, or consigned to the Philippine Islands for sale, including the value of all cartons, cases, crates, boxes, sacks, and coverings of any kind, and all other costs, charges, and expenses incident to placing the merchandise in condition, packed ready for shipment to the Philippine Islands.

(b) Whenever an article is subject to an alternative minimum ad valorem rate, the alternative ad valorem duty shall be ascertained by applying the corresponding ad valorem rate to such merchandise, inclusive of all costs and charges mentioned in clause (a) of this rule.

(c) The term "retail package" wherever used in this act, shall be held to mean any article, goods, wares, or merchandise, together with the holders, containers, packages, or packing, in which such article, goods, wares, or merchandise is usually held, contained, or packed, at the time of its sale to the public in usual retail quantities.

(d) Wherever it is provided in this act that articles, goods, wares, or merchandise shall be dutiable "including weight of immediate containers," the dutiable weight thereof shall be held to be the weight of same, together with the weight of the immediate container, holder, or packing only: *Provided*, That wherever in this act the term "including weight of immediate containers," and the term "retail package" are both used in the same paragraph or clause, the dutiable weight shall be the weight of the retail package.

(e) Wherever it is provided in this act that articles, goods, wares, or merchandise shall be dutiable by "gross weight," the dutiable weight thereof shall be held to be the weight of same, together with the weight of all containers, packages, holders, and packing, of whatsoever kind or character, in which said articles, goods, wares, and merchandise are contained, held, or packed, at the time of importation.

(f) Articles, goods, wares, or merchandise, affixed to cardboard, cards, paper, wood, or similar common material, shall be dutiable together with the weight of such packing.

(g) The usual tapes, boards, and immediate wrapping shall be considered as a part of the dutiable weight of textiles.

(h) No duties shall be assessed on account of the usual coverings or holdings of articles, goods, wares, or merchandise, dutiable otherwise than ad valorem, nor those free of duty, except as in this act expressly provided, but if there be used for covering or holding imported articles, goods, wares, or merchandise, whether dutiable or free, any unusual article, form, or material adapted for use otherwise than in the bona fide transportation of such articles, goods, wares, or merchandise to the Philippine Islands, duty shall be levied and collected on such covering or holding in accordance with the corresponding paragraphs of this act.

(i) Whenever the interior container or packing of any article dutiable by weight is of an unusual character, such as silk-lined cases, cases of fine wood, silk, leather, or imitations thereof, such as are used to contain jewelry, plate, trinkets, and the like, such containers or packing shall be dutiable at the rate applicable to the component material of chief value.

(j) When a single package contains imported merchandise dutiable according to different weights, or weight and ad valorem, the common exterior receptacle shall be prorated and the different proportions thereof treated in accordance with the provisions of this rule as to the dutiability or nondutiability of such packing.

(k) Where articles, goods, wares, or merchandise, dutiable by weight, and not otherwise specially provided for, are customarily contained in packing, packages, or receptacles of uniform or similar character, it shall be the duty of the insular collector of customs from time to time to ascertain by tests the actual weight or quantity of such articles, goods, wares, or merchandise and the actual weight of the packages, packing, or receptacles thereof, respectively, in which the same are customarily imported, and upon such ascertainment to prescribe rules for estimating the dutiable weight or quantity thereof, and thereafter such articles, goods, wares, or merchandise imported in such customary packing, packages, or receptacles shall be entered, and the duties thereon levied and collected, upon the basis of such estimated dutiable weight or quantity: *Provided*, That if the importer, consignee, or agent shall be dissatisfied with such estimated dutiable weight or quantity, and shall file with the collector of customs prior to the delivery of the packages designated for examination a written specification of his objection thereto, or if the collector of customs shall have reason to doubt the exactness of the prescribed weight or quantity in any instance, it shall be his duty to cause such actual weights or quantities to be ascertained.

PROHIBITED IMPORTATIONS.

SEC. 6. That importation into the Philippine Islands of the following articles is prohibited:

(a) Dynamite; gunpowder; similar explosives; firearms and detached parts therefor, except in accordance with enactment of the Philippine legislature.

(b) Books; pamphlets; printed matter; manuscripts; typewritten matter; paintings; illustrations; figures; objects; of obscene or indecent character, or subversive of public order.

(c) Roulette wheels; gambling outfits; loaded dice; marked cards; machines, apparatus, or mechanical devices used in gambling or in the distribution of money, cigars, or other articles, when such distribution is dependent upon chance.

(d) Any article manufactured in whole or in part of gold or silver, or alloys thereof, falsely marked or stamped, in violation of the act of Congress of June 13, 1906, entitled "An act forbidding the importation, exportation, or carriage in interstate commerce of falsely or spuriously stamped articles of merchandise made of gold or silver or their alloys, and for other purposes."

(e) Any article violating the provisions of the act of Congress of June 30, 1906, entitled "An act for preventing the manufacture, sale, or transportation of adulterated or misbranded or poisonous or deleterious foods, drugs, medicines, and liquors, and for regulating traffic therein, and for other purposes," commonly known as "the pure-food law."

(f) Lottery tickets; advertisements thereof; lists of drawings therein; which, after seizure upon illegal entry, shall, together with the proceeds thereof, be forfeited to the government of the Philippine Islands, after due process of law.

(g) Opium, in whatever form, except by the government of the Philippine Islands, and by pharmacists, duly licensed and registered as such, under the laws in force in said islands, and for medicinal purposes only.

(h) Opium pipes; parts therefor; of whatsoever material.

ABBREVIATIONS.

SEC. 7. That the following abbreviations employed in this act shall represent the terms indicated:

Hectog. for hectogram.

Kilo. for kilogram.

Kilos. for kilograms.

Hectol. for hectoliter.

DEFINITIONS.

SEC. 8. The term "pharmaceutical product," wherever used in this act, shall be held to include all medicines or preparations, recognized in the United States Pharmacœpia or National Formulary for internal or external use, and any substance, or mixture of substances, used for the cure, mitigation, or prevention of human or animal diseases, provided the same are not otherwise provided for in this act.

The term "proprietary," as applied to medicinal remedies, wherever used in this act, shall be held to mean a "preparation the manufacture or sale of which is restricted, through patent of the drug or combination of drugs, copyright of the label or name, or in any other manner; or a preparation concerning which the producer or manufacturer claims a private formula."

Wherever in this act the words "the same" appear as the first words of a paragraph, they shall be held to refer to and to mean the same as the caption of the preceding paragraph. Should such words appear as the first words of a clause, they shall be held to refer to and to mean the same as the clause which immediately precedes the one in which they are used.

PAYMENT OF DUTIES.

SEC. 9. That the rates of duty established in this act are stated in money of the United States of America, but that payment thereof shall be made in Philippine currency or its equivalent in money of the United States of America.

METRIC SYSTEM.

SEC. 10. That the metric system of weights and measures, as authorized by sections thirty-five hundred and sixty-nine and thirty-five hundred and seventy of the Revised Statutes of the United States, and at present in use in the Philippine Islands, shall be continued.

The meter is equal to thirty-nine and thirty-seven one hundredths inches.

The liter is equal to one and five hundred and sixty-seven ten-thousandths quarts, wine measure.

The kilogram is equal to two and two thousand and forty-six ten-thousandths pounds, avoirdupois.

RATES OF DUTIES.

SEC. 11. That the rates of duties to be collected on articles, goods, wares, or merchandise imported into the Philippine Islands shall be as follows: *Provided,* That no article bearing evident signs of being for sanitary construction shall pay a higher rate of duty than twenty per centum ad valorem: *And provided further,* That no article shall pay a higher rate of duty than one hundred per centum ad valorem, except and unless the same shall be classified under paragraphs ninety-two, two hundred and sixty-two, two hundred and sixty-three, two hundred and sixty-four, or two hundred and sixty-five, in which event the rate of duty thereby resulting shall be collected, anything in this act to the contrary notwithstanding: *And provided further,* That articles of foreign growth, produce, or manufacture, shall be dutiable upon each importation, even though previously exported from the Philippine Islands, except as otherwise specifically provided in this act.

CLASS I.—STONES, EARTHS, GLASS, AND CERAMIC PRODUCTS.

GROUP 1.—*Stones and earths.*

1. Marble, onyx, jasper, alabaster, and similar fine stones:
    (a) In block, rough, or squared only; marble dust, twenty per centum ad valorem.
    (b) In slabs, plates, or steps, sawed or chiseled, polished or not, but without ornamentation, thirty per centum ad valorem.
    (c) Any of these stones, lettered, further manufactured or decorated, not otherwise provided for, forty per centum ad valorem.
2. Stones, other, natural or artificial, gross weight:
    (a) In block, rough or squared only, one hundred kilos, twenty-five cents.
    (b) Crushed, sawn, hewn, or dressed, polished or not; in slabs, plates, or steps; one hundred kilos, fifty cents.
    (c) Manufactured into articles not otherwise provided for, one hundred kilos, one dollar.
3. Millstones, grindstones, whetstones, oilstones, and hones of all kinds; emery, carborundum and similar wheels for sharpening, dressing, or polishing; including frames and mountings therefor imported therewith, ten per centum ad valorem.
4. Asbestos, and manufactures thereof, not otherwise provided for, fifteen per centum ad valorem.
5. Mica and lava, and manufactures thereof; Welsbach and other similar mantles for lamps; gas-burner tips, twenty-five per centum ad valorem.
6. Earths, gross weight:
    (a) Fire clay; lime; Roman, Portland, and other hydraulic cement; one hundred kilos, six cents.
    (b) Gypsum, pumice, emery, chalk, kaolin (China clay), unmanufactured; other crude earths and clays not otherwise provided for; one hundred kilos, forty cents.
    (c) The same, advanced in value or condition, but not manufactured, one hundred kilos, one dollar.

7. Manufactures of gypsum, gross weight, one hundred kilos, five dollars.

> *Provided,* That no article classified under this paragraph shall pay a less rate of duty than fifty per centum ad valorem.

8. Manufactures of chalk, including billiard chalk, red chalk, and French and tailors' chalk, including weight of immediate containers, one hundred kilos, four dollars.

9. Common clay; cement; in bricks, squares, tiles, and pipes, not otherwise provided for, ten per centum ad valorem.

10. Ceramic tiles, gross weight:
   (a) Varnished, vitrified or not, undecorated, one hundred kilos, forty-five cents.
   (b) Glazed, ornamented or decorated, one hundred kilos, one dollar and twenty cents.

11. Porcelain, bisque, faience, earthenware, stoneware, and other ceramic wares not otherwise provided for:
   (a) In filters and articles bearing evident signs of being for sanitary construction, and parts therefor identifiable as such, ten per centum ad valorem.
   (b) In common bottles, jars, crucibles, cupels, kitchen utensils, and flowerpots, neither gilt, painted, glazed, decorated, nor ornamented, fifteen per centum ad valorem.
   (c) In articles not otherwise provided for, neither gilt, painted, glazed, decorated, nor ornamented, twenty per centum ad valorem.
   (d) In dishes, tableware, or articles not otherwise provided for, glazed or plain-tinted, but neither gilt, painted, decorated, nor ornamented, twenty-five per centum ad valorem.
   (e) In dishes, tableware, or articles not otherwise provided for, painted, gilt, decorated, or ornamented, forty per centum ad valorem.
   (f) Fine decorated wares, in jardinieres, flower stands, vases, and articles for decorative purposes; statuettes, high and bas-relief; Satsuma, Sevres, and similar fine porcelains, decorated or not, fifty per centum ad valorem.

12. Manufactures of earths and clays not otherwise provided for:
   (a) Plain, twenty-five per centum ad valorem.
   (b) Ornamented or decorated, forty per centum ad valorem.

GROUP 2.—*Precious stones, pearls and imitations thereof.*

13. Pearls; precious and semiprecious stones, including jade, tiger-eye, chalcedony, opal and similar stones, not otherwise provided for; cut or uncut, but unmounted and unset, and not further manufactured; diamond dust and bort, fifteen per centum ad valorem.

14. Doublets, and other imitations of precious and of semiprecious stones; imitation pearls, thirty per centum ad valorem.

GROUP 3.—*Glass and manufactures thereof.*

NOTE.—Articles ground or cut only for the purpose of truing them or fitting stoppers, shall not be held to be cut glass.

15. Common hollow glassware:
- (a) In demijohns, carboys, jars, bottles, flasks, and similar receptacles, whether empty, or in use as containers of merchandise dutiable by weight or measure (except in those cases in which the classification of such containers is otherwise specifically provided for), ten per centum ad valorem.
- (b) Siphon bottles, thirty per centum ad valorem.

16. Glass, crystal, and glass imitating crystal:
- (a) In decanters, glasses, tumblers, cups, goblets, saucers, plates, dishes, pitchers, bowls, candlesticks, pillar-lamps, bracket-lamps, and other articles of table service or for lighting; washbowls, wash basins, soap dishes, toothbrush holders and washstand pitchers; neither cut, engraved, painted, enameled, nor gilt, twenty-five per centum ad valorem.
- (b) The same, cut, engraved, painted, enameled, or gilt, fifty per centum ad valorem.

17. Lamp chimneys:
- (a) Neither engraved nor ornamented (except as to such fluting and finishing as may be made in the process of manufacture), twenty-five per centum ad valorem.
- (b) Other, fifty per centum ad valorem.

18. Glass or crystal in plates, slabs, and similar forms:
- (a) Slabs, cones, or prisms, for paving or roofing, gross weight one hundred kilos, one dollar and sixty-five cents.
- (b) Common window glass, neither polished, beveled, engraved, nor annealed, twenty-five per centum ad valorem.
- (c) The same, set in lead, or frosted plain or in design; plate glass, polished, beveled or not, thirty-five per centum ad valorem.
- (d) The same, engraved or enameled, fifty per centum ad valorem.

19. Mirrors of all kinds, framed or mounted (with whatever material) or not, including the value of the frames and mountings, thirty-five per centum ad valorem.

20. Other manufactures of glass:
- (a) In spectacles, eyeglasses, and goggles, also lenses for same, mounted or unmounted, including the value of the mountings, twenty-five per centum ad valorem.
- (b) In flower stands, vases, urns, and similar articles for toilet and decorative purposes, neither cut, painted, enameled, nor gilt, forty per centum ad valorem.
- (c) The same, cut, painted, enameled, or gilt, sixty per centum ad valorem.
- (d) Powdered or crushed glass, twenty-five per centum ad valorem.
- (e) Manufactures not otherwise provided for, in which glass is the component material of chief value, thirty-five per centum ad valorem.

CLASS II.—COAL, SCHISTS, BITUMENS, AND DERIVATIVES THEREOF.

GROUP 1.—*Coal.*

21. Coal and coke, gross weight one thousand kilos, twenty-five cents.

GROUP 2.—*Schists, bitumens, and derivatives thereof.*

In case of doubt as to the classification of crude petroleum, the following rules shall be observed:

1. A sample of two hundred cubic centimeters shall be taken from each fifty cases or less of the same kind of goods in the importation. If the oil is imported in bulk or in tanks, samples of equal quantities shall be taken from each receptacle in which the oil is contained, sufficient to make more than two liters in all after mixing. The contents of the original containers shall be thoroughly mixed before sampling.

2. These samples shall be thoroughly mixed in a large receptacle and when the discharge of the shipment is terminated, two liters shall be taken therefrom and put into separate bottles, which shall be sealed and identified by labels signed by the customs examiner and interested party, and attached thereto. These bottles shall be forwarded to the customs chemical expert for analysis of their contents.

3. Immediately after this operation the goods shall be cleared and the corresponding duty assessed, but the clearance shall not be deemed definite until the result of the analysis is known.

4. The samples must be analyzed within one month, and the interested party shall have the right to be present when the same are opened and analyzed, provided he has made a written request to that effect at the time of identifying the samples by affixing his signature to the labels. He may also appeal to the insular collector of customs from the report of the experts.

5. Should the interested party in his appeal request a new analysis, this shall be made at his expense if the original analysis be confirmed. In contrary case, the expense shall be borne by the Government.

6. The following shall be considered—

(a) As crude and natural petroleum: That imported in the state in which found when extracted from the well, and which has undergone no operation whatever, whereby the natural chemical composition is altered or modified. When gradually and continuously distilled up to a temperature of three hundred degrees Centigrade, this petroleum must leave a residuum exceeding twenty per centum of its primitive weight. Crude oil obtained from the first distillation of slates or schists shall be classified under this head.

(b) As crude oil: That imported in the condition of crude residuum resulting from the first distillation of crude or natural petroleum from which the more volatile constituents have been removed. Such oils must not contain appreciable portions of volatile products which can be removed by distillation at three hundred degrees Centigrade, and they must show by their physical and chemical properties that they have not undergone subsequent rectification.

22. Tars and pitches not otherwise provided for; unrefined tar oils from whatever source; asphaltums; asphalt; asphalt paving blocks; crude petroleum; crude oils of mineral origin; axle grease of all kinds; track oil; carbolineum and similar compounds; gross weight, one hundred kilos, twenty cents.

NOTE.—Though imported under a name referable to this paragraph, impure paraffin, or other similar products, shall be classified under paragraph eighty-five of this act.

23. Mineral oils, refined:
  (a) For illumination, gross weight, one hundred kilos, seventy-five cents.
  (b) For lubrication or solvents, not otherwise provided for; vaseline in bulk; twenty-five per centum ad valorem.
  (c) Vaseline in retail packages, whether proprietary or not (except when compounded with other substances), thirty per centum ad valorem.
  (d) Naphtha, gasoline, benzine, and similar oils for fuel, gross weight, one hundred kilos, seventy-five cents.
  NOTE.—Oils classified under this clause shall have a specific gravity between 0.635 and 0.747, and a boiling point between seventy and one hundred and thirty degrees Centigrade.

24. Mineral oils; their derivatives, compounds, and products; not otherwise provided for, twenty-five per centum ad valorem.

CLASS III.—METALS AND MANUFACTURES THEREOF.

GROUP 1.—*Gold, silver, and platinum; alloys thereof; gold and silver plated articles.*

25. Gold; platinum; alloys thereof:
  (a) In jewelry; plate and goldsmiths' wares not otherwise provided for, hectog., twelve dollars and fifty cents.
  (b) The same, set with pearls or with precious or semiprecious stones, hectog., twenty-five dollars.
  (c) The same, set with doublets, or with imitations of pearls, or of precious or semiprecious stones, hectog., seventeen dollars and fifty cents.
  (d) Articles or manufactures of gold or platinum (except jewelry), composed in part of other materials, in which the component material of chief value is gold or platinum, not otherwise provided for; pellets for use in dentistry; solder and foil; hectog., three dollars.
  Provided, That no article classified under this paragraph shall pay a less rate of duty than twenty-five per centum ad valorem.

26. Silver; alloys thereof:
  (a) In jewelry; plate and silversmiths' wares not otherwise provided for; hectog., one dollar.

(b) The same, set with pearls or with precious or semiprecious stones, hectog., five dollars.

(c) The same, set with doublets, or with imitations of pearls, or of precious or semiprecious stones, hectog., five dollars.

(d) Articles or manufactures of silver (except jewelry), composed in part of other materials, in which the component material of chief value is silver, not otherwise provided for; solder and foil; hectog., forty cents.

Provided, That no article classified under this paragraph shall pay a less rate of duty than twenty-five per centum ad valorem.

27. Gold and silver plated wares:

(a) In jewelry, kilo., two dollars and forty cents.

(b) In lamps not otherwise provided for; picture frames; knives, forks, and spoons; carriage and coffin fittings; saddlery hardware; foil; kilo., sixty cents.

(c) Not otherwise provided for, kilo., two dollars.

Provided, That no article classified under this paragraph shall pay a less rate of duty than twenty-five per centum ad valorem.

GROUP 2.—Cast iron.

NOTE.—Malleable cast iron and manufactures thereof shall be dutiable as wrought iron.

28. Articles of cast iron, painted or not, but not otherwise coated or ornamented, neither polished nor turned, gross weight:

(a) Bars, beams, plates, grates for furnaces, columns and pipes, one hundred kilos, thirty-five cents.

(b) Other, one hundred kilos, seventy-five cents.

29. Other articles of cast iron (except those covered or coated with gold or silver), fifteen per centum ad valorem.

GROUP 3.— Wrought iron and steel.

30. Wrought iron and steel:

(a) In rails, straight or bent, weighing ten kilos or less per lineal meter; crossings and similar track sections, switch rails, and portable tramways, composed of rails weighing ten kilos or less per lineal meter; metal cross-ties, switches, tongues, frogs, fish-plates and chairs therefor, thirty per centum ad valorem.

(b) In rails weighing more than ten kilos per lineal meter, gross weight, one hundred kilos, thirty-five cents.

NOTE.—Bent rails, crossings and similar track sections, switch rails, and portable tramways, composed of rails weighing more than ten kilos per lineal meter; metal cross-ties, switches, tongues, frogs, fish-plates and chairs therefor shall be classified under clause (B) of this paragraph, with a surtax of thirty per centum.

(c) In bars or beams (except of crucible steel), neither cut to measure, perforated, nor riveted or otherwise fastened together; rods, tires and hoops; gross weight, one hundred kilos, forty cents.

Provided, That bars or rods not exceeding fifteen millimeters in diameter and steel known as "bamboo steel,"
classified under clause (C) of this paragraph, shall not pay
a less rate of duty than fifteen per centum ad valorem.

(d) The same, of crucible steel, gross weight one hundred kilos,
two dollars and sixty-five cents.

Provided, That no article classified under clause (D) of
this paragraph shall pay a less rate of duty than fifteen per
centum ad valorem.

31. Wrought iron or steel in sheets, gross weight:

(a) Plain and unpolished, one hundred kilos, fifty cents.

(b) Polished, corrugated, perforated or cold rolled, galvanized
or not; hoop iron; one hundred kilos, one dollar.

Provided, That any of the articles or materials classified
under this clause, made up in hoops, ridgings, eaves, drain
pipes, gutters, ceilings, shingles, ceiling centers, borders,
friezes, dados, and similar articles, shall be dutiable at the
rate herein provided, with a surtax of one hundred per
centum.

(c) Tinned; terne-plate; tin-plate; one hundred kilos, one dollar
and twenty cents.

32. Wrought iron or steel, in pieces, in the rough, gross weight:

(a) Neither polished, turned nor adjusted, one hundred kilos,
sixty-five cents.

(b) Rough-turned or lathed, but neither polished nor adjusted,
one hundred kilos, one dollar.

33. Wrought iron or steel, in pieces, finished, gross weight:

(a) Wheels weighing each more than one hundred kilos; axles,
springs, brake-shoes, drawbars, brake-beams, bumpers,
couplings, lubricating boxes, and similar articles for railways and tramways; one hundred kilos, forty-five cents.

(b) Wheels weighing each one hundred kilos or less; axles and
springs for vehicles, not otherwise provided for; one hundred kilos, one dollar and five cents.

34. Wrought iron or steel in large pieces, composed of bars, beams,
or sheets, for structural purposes, perforated or cut to measure, fastened together or not, gross weight, one hundred
kilos, one dollar and twenty-five cents.

35. Wrought iron or steel pipes, gross weight:

(a) Plain, painted, tarred, or galvanized, one hundred kilos,
one dollar and ten cents.

(b) Other (except those coated or covered with gold or silver),
one hundred kilos, one dollar and fifty cents.

36. Wrought iron or steel wire:

(a) More than one millimeter in diameter, plain, galvanized,
or coppered; wire cables and ropes; barbed wire; ten per
centum ad valorem.

(b) One millimeter or less in diameter, plain, galvanized, or
coppered; wire netting; fifteen per centum ad valorem.

(c) Other, including those covered with textiles, twenty-five
per centum ad valorem.

(d) Gauze, cloths and screenings, in the piece, twenty per
centum ad valorem.

(e) In other manufactures (except those covered or coated with gold or silver), not otherwise provided for, twenty-five per centum ad valorem.

37. Wrought iron or steel chains, in the piece or otherwise (except in trinkets or jewelry):
   (a) Exceeding five millimeters in diameter, ten per centum ad valorem.
   (b) Other, plain, painted, or galvanized, fifteen per centum ad valorem.
   (c) The same, covered or coated with other metals (except gold or silver), twenty-five per centum ad valorem.

38. Anvils, ten per centum ad valorem.

39. Nuts, bolts, rivets, and washers, one hundred kilos, two dollars.

40. Nails, clasp nails, and staples, ten per centum ad valorem.

41. Screws, tacks, and brads, fifteen per centum ad valorem.

42. Saddlery hardware (except chains and buckles), plain, or covered or coated with other metals or materials (except gold or silver), fifteen per centum ad valorem.

43. Buckles (except trinkets or ornaments, or covered or coated with gold or silver), fifteen per centum ad valorem.

44. Cutlery:
   (a) Butchers', shoemakers', saddlers', plumbers', painters', pruning, budding, kitchen, bread, and cheese knives; table knives and forks, with handles of common wood, or of iron, japanned or not, not covered or coated with other metals; common scissors or shears, plain, glazed, or japanned; grass, garden, hedge, pruning, and sheep shears; fishhooks; twenty per centum ad valorem.
   (b) Pocket cutlery; hunting and sheath knives; side arms (not fire) and parts therefor; razors; other cutlery, including scissors and shears not otherwise provided for (except those covered or coated with gold or silver), thirty per centum ad valorem.
   (c) Sword canes and similar articles and weapons with concealed blades, eighty per centum ad valorem.

45. Firearms of all kinds and detached parts therefor, forty per centum ad valorem.

46. Manufactures of terne-plate or tin-plate:
   (a) In articles not otherwise p v d d for, plain, painted, varnished, or japanned, fifteen per centum ad valorem.
   (b) The same, including vehicle lamps, covered, coated or combined with other metals or materials (except gold or silver), twenty per centum ad valorem.
   (c) Vehicle lamps, covered or coated to any extent with gold or silver, in which the component material of chief value is tin-plate, twenty-five per centum ad valorem.

47. Manufactures not otherwise provided for, in which wrought iron or steel is the component material of chief value:
   (a) Plain, painted, varnished, or japanned, or covered or coated with lead, tin, or zinc, fifteen per centum ad valorem.
   (b) Other (except those covered or coated with gold or silver), twenty per centum ad valorem.

GROUP 4.—*Copper and alloys thereof.*

48. Copper; alloys thereof; in bars, pipes, and sheets; alloys of copper in lumps and ingots (except Muntz metal); ten per centum ad valorem.
49. Copper; alloys thereof; in wire:
    (a) Plain, fifteen per centum ad valorem.
    (b) Blanched, gilt or nickeled, twenty-five per centum ad valorem.
    (c) Covered with textiles, not otherwise provided for, or with insulating materials; cables for conducting electricity; trolley wire; ten per centum ad valorem.
    (d) Covered with silk, not otherwise provided for, twenty-five per centum ad valorem.
    (e) Gauze, cloths and screenings, in the piece, twenty per centum ad valorem.
    (f) Manufactures not otherwise provided for, in which wire of copper or its alloys is the component material of chief value (except when covered or coated with gold or silver), twenty-five per centum ad valorem.
50. Manufactures not otherwise provided for, in which copper or alloys thereof is the component material of chief value:
    (a) Plain, polished, varnished, painted, tinned, or japanned, twenty per centum ad valorem.
    (b) Other (except those covered or coated with gold or silver), twenty-five per centum ad valorem.

GROUP 5.—*Other metals and alloys thereof.*

51. Mercury, gross weight, kilo, ten cents.
52. Nickel, aluminum, and alloys thereof:
    (a) In bars, sheets, pipes, and wire, fifteen per centum ad valorem.
    (b) In articles not otherwise provided for, twenty-five per centum ad valorem.
53. Tin and alloys thereof:
    (a) In bars, sheets, pipes, and wire; in thin leaves (tin foil) alloys in lumps or ingots; ten per centum ad valorem.
    (b) In articles not otherwise provided for (except those covered or coated with gold or silver), twenty-five per centum ad valorem.
54. Zinc, lead, and metals not otherwise provided for; alloys thereof:
    (a) In bars, sheets, pipes, wire, and type; sanitary traps, and other plain articles bearing evident signs of being for sanitary construction; alloys in lumps or ingots; ten per centum ad valorem.
    (b) In plain articles not otherwise provided for, fifteen per centum ad valorem.
    (c) In articles gilt, nickeled or otherwise embellished (except those covered or coated with gold or silver), twenty-five per centum ad valorem.

CLASS IV.—SUBSTANCES EMPLOYED IN PHARMACY, AND CHEMICAL INDUSTRIES; DRUGS, CHEMICALS, PIGMENTS, AND VARNISHES.

GROUP 1.—*Simple drugs.*

55. Oleaginous seeds, copra, and cocoanuts, gross weight:
    (a) Crude, one hundred kilos, eighty cents.
    (b) In meal, flour or cakes, not otherwise provided for, one hundred kilos, one dollar and fifty cents.
56. Resins and gums:
    (a) Colophony (common or navy resin); Burgundy and similar pitch; Stockholm tar; ten per centum ad valorem.
    (b) Other, when not in the form of a pharmaceutical product or preparation, twenty per centum ad valorem.
57. Drugs, such as barks, beans, berries, buds, bulbs, bulbous roots, fruits, flowers, dried fibers, grains, herbs, leaves, lichens, mosses, stems, seeds aromatic and seeds of morbid growth, weeds, woods, and similar vegetable products, crude, neither edible nor in the form of a pharmaceutical product or preparation, not otherwise provided for, including weight of immediate containers, one hundred kilos, three dollars.
    Provided, That no article classified under this paragraph shall pay a less rate of duty than twenty-five per centum ad valorem.
58. Ginseng root, kilo, five dollars.
    Provided, That no article classified under this paragraph shall pay a less rate of duty than twenty-five per centum ad valorem.
59. Animal products employed in medicine, crude, neither edible nor in the form of a pharmaceutical product or preparation, not otherwise provided for, including weight of immediate containers, one hundred kilos, four dollars.
    Provided, That no article classified under this paragraph shall pay a less rate of duty than twenty-five per centum ad valorem.

GROUP 2.—*Pigments, paints, dyes, and varnishes.*

60. Mineral pigments of common, natural occurrence, such as ochers, haemitites, barytes and manganese; substances prepared for calcimines and whitewash; dry; ten per centum ad valorem.
    NOTE.—Any substance otherwise subject to classification under this paragraph shall, when imported in the form of a liquid or paste, be dutiable under clause (d) of paragraph sixty-one.
61. Pigments and paints not otherwise provided for:
    (a) White or red lead, dry, fifteen per centum ad valorem.
    (b) The same, in liquid or paste; putty of all kinds; bituminous paints made from mineral pitch or coal tar (not aniline dyes), twenty per centum ad valorem.
    (c) Pigments not otherwise provided for, dry, twenty per centum ad valorem.
    (d) The same, in liquid or paste, twenty-five per centum ad valorem.

62. Varnishes and wood fillers of all kinds, fifteen per centum ad valorem.
63. Spirits of turpentine, ten per centum ad valorem.
64. Inks:
    (a) Printing and lithographic, in any form, fifteen per centum ad valorem.
    (b) Other, in any form, twenty-five per centum ad valorem.
65. Pencils of paper or wood, filled with lead or other materials; pencils of lead; charcoal and other crayons not otherwise provided for, fifteen per centum ad valorem.
66. Dyes and dyestuffs; tan bark and tanning extracts; not otherwise provided for:
    (a) Woods, barks, roots, and similar natural products, for dyeing or tanning, ten per centum ad valorem.
    (b) Extracts from the same. for dyeing or tanning; cutch in any form; fifteen per centum ad valorem.
    (c) Cochineal; indigo (natural or synthetic); colors derived from coal; chemical dye colors not otherwise provided for; thirty per centum ad valorem.
67. Graphite and manufactures of the same (except axle grease); polishing, dressing, cleansing, and preserving preparations for shoes and leather; twenty-five per centum ad valorem.

GROUP 3.—*Chemical and pharmaceutical products.*

68. Sulphur, gross weight, one hundred kilos, fifty cents.
69. Bromine, boron, iodine, and phosphorus, twenty per centum ad valorem.
70. Inorganic acids:
    (a) Hydrochloric, boric, nitric, and sulphuric; mixtures of two or more of the same; gross weight, one hundred kilos, thirty-five cents.
    (b) Carbon dioxide (liquid carbonic acid); sulphur dioxide; twenty per centum ad valorem.
    (c) Not otherwise provided for, twenty-five per centum ad valorem.
71. Organic acids, not otherwise provided for:
    (a) Carbolic, ten per centum ad valorem.
    (b) Other, twenty-five per centum ad valorem.
72. Oxides and hydroxides of potassium, sodium, barium, and other caustic alkalies, not otherwise provided for; soda ash; gross weight, one hundred kilos, fifty cents.
73. Aqua ammonia; anhydrous ammonia: fifteen per centum ad valorem.
74. Inorganic salts:
    (a) Sulphates of ammonium and potassium; chloride of potassium; phosphates and superphosphates of lime; nitrates of potassium and sodium; other chemical and artificial fertilizers; five per centum ad valorem.
    (b) Calcium hypochlorite (chloride of lime), ten per centum ad valorem.
    (c) Not otherwise provided for, twenty-five per centum ad valorem.

75. Organic salts not otherwise provided for, twenty-five per centum
    ad valorem.
    NOTE.—No acids or double salts shall be dutiable under
    this paragraph.
76. Mixtures of denaturants; formálin; potassium bi-tartrate (cream
    of tartar, argols, wine lees); ten per centum ad valorem.
77. Chemical products, compounds, and elements, not otherwise pro-
    vided for, twenty-five per centum ad valorem.
78. Alkaloids and their salts (except those of opium or of cinchona
    bark); salts of gold, silver, and platinum; thirty-five per
    centum ad valorem.
79. Opium in any form, and preparations thereof, for medicinal pur-
    poses, not otherwise provided for, subject to the provisions of
    section six of this act, thirty-five per centum ad valorem.
80. Proprietary and patent medicinal fixtures and compounds; Chi-
    nese and similar medicines:
    (a) Without alcohol, or containing not to exceed fourteen per
        centum of alcohol, fifty per centum ad valorem.
    (b) Containing more than fourteen per centum of alcohol,
        seventy-five per centum ad valorem.
81. Pharmaceutical products; medicinal preparations; plasters and
    poultices; capsules, empty; not otherwise provided for;
    thirty per centum ad valorem.
82. Aseptic and antiseptic surgical dressings, including absorbent cot-
    ton, medicated or not; catgut, silk, and similar ligatures for
    use in surgery or dentistry; fifteen per centum ad valorem.

GROUP 4.—*Oils, fats, waxes, and derivatives thereof.*

83. Fixed vegetable oils, solid or liquid:
    (a) In receptacles weighing each (contents included) more than
        two kilos, fifteen per centum ad valorem.
    (b) In other receptacles, proprietary or not (except when com_
        pounded with other substances, or in capsules), twenty_
        five per centum ad valorem.
84. Animal oils and fats, not otherwise provided for:
    (a) Crude, ten per centum ad valorem.
    (b) Refined, in receptacles weighing each (contents included)
        more than two kilos, fifteen per centum ad valorem.
    (c) The same, in other receptacles, proprietary or not (except
        when compounded with other substances, or in capsules),
        twenty-five per centum ad valorem.
85. Mineral, vegetable, and animal wax; paraffin and stearin:
    (a) Crude, ten per centum ad valorem.
    (b) In candles, twenty per centum ad valorem.
    (c) In manufactures not otherwise provided for, thirty per
        centum ad valorem.
86. Soaps; soap powders; similar cleansing and scouring preparations
    or compositions; not otherwise provided for; fifteen per
    centum ad valorem.
87. Essential oils; perfumery, and products used in the manufacture
    thereof; toilet preparations:
    (a) Essential oils, natural or artificial, fifty per centum ad
        valorem.

(b) Perfumery; products used in the manufacture thereof; toilet preparations, including powders, oils, cosmetics, hair dyes, tooth soaps and tooth powders, grease paints, and similar articles for toilet purposes; not otherwise provided for; incense; joss sticks; forty per centum ad valorem.

GROUP 5.— *Various.*

88. Bone char, suitable for use in decolorizing sugar, ten per centum ad valorem.
89. Starch; fecula; dextrin; for industrial purposes; gross weight, one hundred kilos, two dollars.
90. Glues; albumens; gelatins; isinglass; manufactures thereof; twenty-five per centum ad valorem.
91. Explosives:
    (a) Dynamite, giant and blasting powder, and similar explosives; miners' fuses and caps; explosive signals; ten per centum ad valorem.
    (b) Other; fire works; cartridges, fixed ammunition, primers and percussion caps, for firearms; thirty per centum ad valorem.
    (c) Fire crackers and toy torpedoes, including weight of immediate containers, kilo, twenty cents.
92. Matches of all kinds, including weight of immediate containers, kilo, twenty cents.

CLASS V.—COTTON AND MANUFACTURES THEREOF.

GROUP 1.—*Cotton waste.*

93. Cotton waste, ten per centum ad valorem.

GROUP 2.— *Yarns, threads, and cordage.*

94. Yarns, not otherwise provided for, in hanks, cops, or bobbins, fifteen per centum ad valorem.
95. Yarns or threads for sewing, crocheting, darning, or embroidering; mercerized yarns or threads; twenty-five per centum ad valorem.
96. Threads or twines for sewing sails and sacks; rope and cordage; fishing nets; wicks for making candles and matches; twenty per centum ad valorem.
97. Hammocks; tennis nets; manufactures of netting not otherwise provided for; forty per centum ad valorem.
98. Felts; batting; mops and swabs of cotton yarns; fifteen per centum ad valorem.

GROUP 3.—*Textiles.*

NOTE.—When textiles, included in this group, contain an admixture of materials, are broché, embroidered, trimmed, or made-up, they shall be subject to the corresponding surtaxes prescribed in general rules two to eleven, inclusive. Textiles woven with a colored yarn on the selvage, or with a colored selvage stripe not exceeding ten millimeters in width, shall not be considered as manufactured with dyed yarns.

99. Textiles, plain and without figures, napped or not, weighing eight kilos or more per one hundred square meters, having:
   (a) Up to eighteen threads, kilo, ten cents.
   (b) From nineteen to thirty-one threads, kilo, fourteen cents.
   (c) From thirty-two to thirty-eight threads, kilo, twenty cents.
   (d) From thirty-nine to forty-four threads, kilo, twenty-six cents.
   (e) Forty-five threads or more, kilo, thirty-two cents.
      Provided, That any textile classified under this paragraph, stamped, printed, or manufactured with dyed yarns, shall be dutiable as such, with a surtax of thirty per centum; and
      Provided further, That no embroidered textile classified under this paragraph shall pay a less rate of duty than twenty-five per centum ad valorem, and any embroidered textile so classified shall be subject to all of the surtaxes applicable thereto under this act, computed upon the ascertained amount of duty, whether the rate found applicable shall be specific or ad valorem.

100. The same, weighing less than eight kilos per one hundred square meters, having:
   (a) Up to eighteen threads, kilo, eighteen cents.
   (b) From nineteen to thirty-one threads, kilo, twenty-seven cents.
   (c) From thirty-two to thirty-eight threads, kilo, thirty-four cents.
   (d) From thirty-nine to forty-four threads, kilo, forty cents.
   (e) Forty-five threads or more, kilo, fifty cents.
      Provided, That any textile classified under this paragraph, stamped, printed, or manufactured with dyed yarns, shall be dutiable as such, with a surtax of forty per centum; and
      Provided further, That no embroidered textile classified under this paragraph shall pay a less rate of duty than twenty-five per centum ad valorem, and any embroidered textile so classified shall be subject to all of the surtaxes applicable thereto under this act, computed upon the ascertained amount of duty, whether the rate found applicable shall be specific or ad valorem.

101. Textiles, twilled or figured in the loom, napped or not, weighing ten kilos or more per one hundred square meters, having:
   (a) Up to eighteen threads, kilo, fourteen cents.
   (b) From nineteen to thirty-one threads, kilo, eighteen cents.
   (c) From thirty-two to thirty-eight threads, kilo, twenty-four cents.
   (d) Thirty-nine to forty-four threads, kilo, thirty cents.
   (e) Forty-five threads or more, kilo, thirty-four cents.
      Provided, That any textile classified under this paragraph, stamped, printed, or manufactured with dyed yarns, shall be dutiable as such, with a surtax of thirty per centum; and
      Provided further, That no embroidered textile classified under this paragraph shall pay a less rate of duty than twenty-five per centum ad valorem, and any embroidered

textile so classified shall be subject to all of the surtaxes applicable thereto under this act, computed upon the ascertained amount of duty, whether the rate found applicable shall be specific or ad valorem.

102. The same, weighing less than ten kilos per one hundred square meters, having:

    (a) Up to eighteen threads, kilo, twenty-four cents.

    (b) From nineteen to thirty-one threads, kilo, thirty-two cents.

    (c) From thirty-two to thirty-eight threads, kilo, forty-two cents.

    (d) From thirty-nine to forty-four threads, kilo, fifty-two cents.

    (e) Forty-five threads or more, kilo, sixty cents.

    Provided, That any textile classified under this paragraph, stamped, printed, or manufactured with dyed yarns, shall be dutiable as such, with a surtax of forty per centum; and

    Provided further, That no embroidered textile classified under this paragraph shall pay a less rate of duty than twenty-five per centum ad valorem, and any embroidered textile so classified shall be subject to all of the surtaxes applicable thereto under this act, computed upon the ascertained amount of duty, whether the rate found applicable shall be specific or ad valorem.

103. Piqués of all kinds, kilo, thirty-eight cents.

    Provided, That any article classified under this paragraph shall not pay a less rate of duty than thirty per centum ad valorem.

104. Cotton blankets:

    (a) In the piece, single or double, but neither hemmed nor bound, kilo, ten cents.

    (b) Stamped, printed, or manufactured with dyed yarns, kilo, thirteen cents.

    Provided, That any blanket classified under this paragraph, hemmed or bound, shall be dutiable as such, with a surtax of thirty per centum.

105. Plushes, velvets, velveteens, and other pile fabrics (except in towels and bath robes) (see rule six), kilo, fifty cents.

106. Bath robes and towels manufactured with pile warp, twenty-five per centum ad valorem.

107. Knitted goods (see rule six):

    (a) In the piece, twenty per centum ad valorem.

    (b) In jerseys, undershirts, drawers, stockings, or socks, twenty-five per centum ad valorem.

    (c) In other articles, thirty-five per centum ad valorem.

    Provided, That any article classified under this paragraph, embroidered, shall be dutiable as such, with a surtax of thirty per centum, computed upon the ascertained amount of duty under the corresponding clause thereof.

108. Tulles (see rule six):

    (a) Plain, kilo, fifty-six cents.

    (b) Figured or embroidered on the loom, kilo, one dollar and three cents.

    Provided, That no article classified under this paragraph shall pay a less rate of duty than thirty per centum ad valorem; and

Provided further, That any of the same embroidered or figured after weaving, out of the loom, shall be dutiable according to the respective clause, with a surtax of sixty per centum; and

Provided further, That if the embroidery consists of metal threads, the surtax shall be eighty per centum; and

Provided further, That these surtaxes shall be computed upon the ascertained amount of duty, whether the rate found applicable be specific or ad valorem.

109. Laces and blondes (see rule six):
  (a) Lace curtains, bedspreads, pillow shams, and bed sets, unhemmed, hemmed, or bound, made on the Nottingham lace-curtain or warp machines, kilo, fifty cents.
  (b) Other, sixty per centum ad valorem.

110. Carpeting, thirty per centum ad valorem.

111. Textiles called tapestries:
  (a) In the piece, kilo, twenty cents.
  (b) In made up articles, kilo, thirty cents.
  Provided, That no article classified under this paragraph shall pay a less rate of duty than forty per centum ad valorem.

112. Wicks for lamps, including weight of immediate containers, kilo, fifteen cents.

113. Trimmings, ribbons, braids, tape, and galloons, including weight of immediate containers (see rule seven):
  (a) Tape: boot straps; kilo, twenty cents.
  (b) Other, kilo, fifty cents.
  Provided, That no article classified under clause (b) of this paragraph shall pay a less rate of duty than thirty per centum ad valorem.

114. Shoe and corset laces, including weight of immediate containers, kilo, thirty-five cents.

115. Cinches, saddle girths, reins, halters, and bridles, twenty-five per centum ad valorem.

116. Ribbons or bands for the manufacture of any of the articles enumerated in paragraph one hundred and fifteen, fifteen per centum ad valorem.

117. Waterproof or caoutchouc stuffs in combination with cotton textiles; cotton elastic textiles manufactured with threads or gum elastic; manufactures thereof; twenty-five per centum ad valorem.

118. Manufactures of cotton, not otherwise provided for, twenty-five per centum ad valorem.

CLASS VI.—MANUFACTURES OF HEMP, FLAX, ALOE, JUTE, AND VEGETABLE FIBERS, NOT OTHERWISE PROVIDED FOR.

GROUP 1.—*Yarns, threads, and cordage.*

119. Yarns, not otherwise provided for, fifteen per centum ad valorem.

120. Threads, twines, ropes, cordage, and manufactures thereof:
  (a) Twines, rope-yarns, ropes, and cordage, exceeding fifteen grams in weight per each ten meters; fishing nets; twenty per centum ad valorem.

    (b) Threads, twines, cords, and yarns twisted, weighing more than five and not exceeding fifteen grams per each ten meters, twenty-five per centum ad valorem.

    (c) The same, weighing five or less grams per each ten meters, thirty per centum ad valorem.

    (d) Hammocks; tennis nets; manufactures of netting not otherwise provided for; forty per centum ad valorem.

121. Gunny sacks, each, two cents.

Group 2.—*Textiles.*

    Note.—When textiles, included in this group, contain an admixture of materials, are embroidered, trimmed, or made up, they shall be subject to the corresponding surtax prescribed in general rule two to eleven, inclusive.

    Textiles woven with a colored yarn on the selvage, or with a colored selvage stripe not exceeding ten millimeters in width, shall not be considered as manufactured with dyed yarns.

122. Textiles of hemp, flax, aloe, jute, and vegetable fibers, not otherwise provided for, plain, twilled, or damasked, weighing thirty-five kilos or more per one hundred square meters, having:

    (a) Up to ten threads, used for bagging and baling, weighing forty-five kilos or more per one hundred square meters, kilo, one cent.

    (b) The same, weighing from thirty-five to forty-five kilos per one hundred square meters, kilo, two cents.

    (c) Up to ten threads, for other purposes, kilo, seven cents.

    (d) From eleven to eighteen threads, kilo, ten cents.

    (e) Nineteen threads or more, kilo, fifteen cents.

    Provided, That any textile classified under this paragraph, bleached, half bleached, stamped, or printed, shall be dutiable as such, with a surtax of fifteen per centum; and

    Provided further, That any textile classified under this paragraph manufactured with dyed yarns shall be dutiable as such, with a surtax of twenty-five per centum.

123. The same, weighing from twenty to thirty-five kilos per one hundred square meters, having:

    (a) Up to ten threads, used for bagging and baling, kilo, two cents.

    (b) Up to ten threads, for other purposes, kilo, nine cents.

    (c) From eleven to eighteen threads, kilo, fourteen cents.

    (d) From nineteen to twenty-four threads, kilo, eighteen cents.

    (e) From twenty-five to thirty threads, kilo, twenty-two cents.

    (f) From thirty-one to thirty-eight threads, kilo, thirty cents.

    (g) Thirty-nine threads or more, kilo, forty cents.

    Provided, That any textile classified under this paragraph, bleached, half bleached, stamped, or printed, shall be dutiable as such, with a surtax of twenty-five per centum; and

    Provided further, That any textile classified under this paragraph, manufactured with dyed yarns, shall be dutiable as such, with a surtax of forty per centum.

124. The same, weighing from ten to twenty kilos per one hundred square meters, having:

(a) Up to eighteen threads, kilo, twelve cents.
(b) From nineteen to twenty-four threads, kilo, twenty cents.
(c) From twenty-five to thirty threads, kilo, twenty-eight cents.
(d) From thirty-one to thirty-eight threads, kilo, thirty-six cents.
(e) Thirty-nine threads or more, kilo, fifty-six cents.

Provided, That any textile classified under this paragraph, bleached, half bleached, stamped, or printed, shall be dutiable as such, with a surtax of thirty per centum; and

Provided further, That any textile classified under this paragraph, manufactured with dyed yarns, shall be dutiable as such, with a surtax of fifty per centum; and

Provided further, That no article classified under this paragraph shall pay a less rate of duty than twenty per centum ad valorem.

125. The same, weighing less than ten kilos per one hundred square meters, having:
(a) Up to twelve threads, kilo, eighteen cents.
(b) From thirteen to twenty-two threads, kilo, thirty-two cents.
(c) From twenty-three to thirty threads, kilo, forty-five cents.
(d) From thirty-one to thirty-eight threads, kilo, fifty-six cents.
(e) Thirty-nine threads or more, kilo, ninety cents.

Provided, That any textile classified under this paragraph, bleached, half bleached, stamped, or printed, shall be dutiable as such, with a surtax of thirty per centum; and

Provided further, That any textile classified under this paragraph, manufactured with dyed yarns, shall be dutiable as such, with a surtax of fifty per centum; and

Provided further, That no article classified under this paragraph shall pay a less rate of duty than twenty per centum ad valorem.

126. Plushes, velvets, velveteens, and other pile fabrics (see Rule six), thirty per centum ad valorem.
127. Knitted goods (see Rule six):
(a) In the piece, made up into jerseys, undershirts, drawers, stockings, or socks, thirty per centum ad valorem.
(b) In other articles, forty per centum ad valorem.
128. Tulles and laces (see Rule six), sixty per centum ad valorem.
129. Carpeting, thirty-five per centum ad valorem.
130. Tapestries, kilo, forty cents.

Provided, That no article classified under this paragraph shall pay a less rate of duty than fifty per centum ad valorem.

131. Trimmings, ribbons, braid, tape, and galloons, including weight of immediate containers (see rule seven):
(a) Tape; boot straps; kilo, thirty cents.
(b) Other, kilo, sixty cents.

Provided, That no article classified under clause (b) of this paragraph shall pay a less rate of duty than thirty-five per centum ad valorem.

132. Shoe and corset laces, including weight of immediate containers, kilo, forty cents.
133. Cinches, saddle girths, reins, halters, and bridles, thirty-five per centum ad valorem.
134. Ribbons or bands for the manufacture of any of the articles enumerated in paragraph one hundred and thirty-three, twenty per centum ad valorem.
135. Waterproof or caoutchouc stuffs in combination with textiles of vegetable fibers (other than cotton); elastic textiles of any of the same manufactured with threads of gum elastic; manufactures thereof; thirty per centum ad valorem.
136. Manufacturers of vegetable fibers, not otherwise provided for, thirty per centum ad valorem.

CLASS VII.—WOOL, BRISTLES, HAIR, AND MANUFACTURES THEREOF.

GROUP 1.—*Unmanufactured.*

137. Wool, not otherwise provided for:
    (a) Combed, prepared for yarns; wool waste; ten per centum ad valorem.
    (b) Combed, and carded or dyed, fifteen per centum ad valorem.

GROUP 2.—*Yarns.*

138. Yarns, thirty per centum ad valorem.

GROUP 3.—*Manufactures.*

139. Bristles; animal hair; manufactures thereof; not otherwise provided for; thirty per centum ad valorem.
140. Human hair, made up into articles or not, fifty per centum ad valorem.
141. Knitted goods (see Rule six):
    (a) In the piece, thirty per centum ad valorem.
    (b) In jerseys, undershirts, drawers, stockings, or socks, thirty-five per centum ad valorem.
    (c) In other articles, forty per centum ad valorem.
142. Textiles of wool, in the piece, thirty-five per centum ad valorem.
143. Manufactures of wool, not otherwise provided for, forty per centum ad valorem.

CLASS VIII.—SILK AND MANUFACTURES THEREOF.

GROUP 1.—*Raw and spun.*

144. Silk waste, twenty-five per centum ad valorem.
145. Spun silks, not twisted, including weight of immediate containers, kilo, one dollar and fifty cents.
146. Floss and twisted silks, thirty-five per centum ad valorem.

GROUP 2.—*Textiles.*

147. Silk, in the Piece, forty per centum ad valorem.
148. Manufactures in which silk, artificial silk, or imitation silk, is the component material of chief value, not otherwise provided for, fifty per centum ad valorem.

CLASS IX.—PAPER AND MANUFACTURES THEREOF.

149. Printing paper, white or colored, suitable for books or newspapers, not printed or otherwise elaborated; sand, glass, emery, carborundum, and similar papers; sheathing and roofing paper; ten per centum ad valorem.

150. Paper, pasteboard, cardboard, bristolboard, strawboard, and pulpboard, white or colored, not otherwise provided for:
    (a) Not printed or otherwise elaborated; writing paper, plain, ruled or padded, but not printed; fifteen per centum ad valorem.
    (b) The same manufactured into articles, including confetti and serpentine; envelopes of all kinds, without printing; twenty per centum ad valorem.

151. Paper of all kinds, pasteboard, cardboard, bristolboard, strawboard, and pulpboard:
    (a) Ruled, printed, engraved, lithographed, surface coated, etched, embossed, or otherwise elaborated; printed or lithographed music, bound or in sheets, with or without words, not otherwise provided for; twenty per centum ad valorem.
    (b) The same, manufactured into articles, not otherwise provided for, twenty-five per centum ad valorem.

152. Cigarette paper, printed or not, fifteen per centum ad valorem.

153. Blank books, ruled or unruled, with printing or not; copying books; twenty per centum ad valorem.

154. Printed books, bound or not, not otherwise provided for, ten per centum ad valorem.

155. Books and albums of lithographs, engravings, etchings, photographs, maps, or charts, not otherwise provided for; painted designs, pastels, and ink drawings, made by hand, for use in manufacturing and in the industrial arts and sciences; thirty per centum ad valorem.

    NOTE.—This paragraph shall not apply to works of art introduced for use as such, even when imported for sale, which shall be classified under paragraph three hundred and twenty-six.

156. Papier maché; carton pierre; indurated pulp or fiber:
    (a) Not further manufactured than in sheets or blocks, ten per centum ad valorem.
    (b) Further manufactured, twenty per centum ad valorem.

CLASS X.—WOOD AND OTHER MATERIALS, AND MANUFACTURES THEREOF.

GROUP 1.— *Wood*.

157. Common wood, including cedar of all kinds:
    (a) In logs or poles, or not further advanced in manufacture than hewn or sawn into rough boards or timber, cubic meter, one dollar.
    (b) Plain, dovetailed, or cut to size, including shingles, laths and fencing, fifteen per centum ad valorem.

158. Fine wood:
    (a) In logs or poles, or not further advanced in manufacture than hewn or sawn into rough boards or timber, twenty per centum ad valorem.
    (b) Planed, dovetailed, or cut to size, twenty-five per centum ad valorem.
159. Wood shavings; sawdust; excelsior; (except those of dye and scented woods); ten per centum ad valorem.
160. Shooks; staves; headings; hoops; bungs; ten per centum ad valorem.
161. Tuns, pipes, casks, and similar receptacles, whether empty or in use as containers of merchandise dutiable by weight or measure (except in those cases in which the classification of such containers is otherwise specifically provided for):
    (a) Suitable for use as containers of liquids, twenty per centum ad valorem.
    (b) Other, ten per centum ad valorem.

GROUP 2.— *Manufactures of wood.*

162. Manufactures of common wood, not otherwise provided for, whether finished, turned, painted, varnished, or not, but neither inlaid, veneered, carved, nor upholstered, nor covered nor lined with stuffs or leather; Vienna or bent-wood furniture; twenty-five per centum ad valorem.
163. Manufactures of fine wood, not otherwise provided for, whether turned, painted, varnished, or polished, or upholstered, covered, or lined with stuffs (except silk or leather), or not; manufactures of common wood, not otherwise provided for, veneered with fine wood, or upholstered, covered, or lined with stuffs (except silk or leather); thirty per centum ad valorem.
164. Manufactures of common or fine wood, not otherwise provided for, gilt, inlaid, veneered with metal, or ornamented with metal or carving, or upholstered, covered, or lined with silk or leather, thirty-five per centum ad valorem.
165. Barbers' and dentists' chairs, of whatever material, twenty-five per centum ad valorem.
166. Bowling alleys; billiard, pool, bagatelle, and similar tables; parts and appurtenances therefor, including balls of whatever material (except chalk and cloth); forty per centum ad valorem.

GROUP 3.— *Various.*

167. Charcoal; firewood; other vegetable fuels; gross weight; one hundred kilos, five cents.
168. Cork:
    (a) Rough or in boards, five per centum ad valorem.
    (b) In stoppers for receptacles, fifteen per centum ad valorem.
    (c) In other articles, twenty-five per centum ad valorem.
169. Straw for manufacturing purposes; rushes; vegetable hair; genista; osiers; bamboo; broomcorn; rattan; reeds; piths; not otherwise provided for:

(a) Crude, or not further advanced in manufacture than cut into straight lengths suitable for sticks for umbrellas, parasols, sun shades, whips, fishing rods, or walking canes, ten per centum ad valorem.
(b) Manufactured into furniture, twenty-five per centum ad valorem.
(c) Manufactured into articles not otherwise provided for, thirty-five per centum ad valorem.
(d) Rattan, split or stripped, bleached or not, twenty per centum ad valorem.

CLASS XI.—ANIMALS, AND ANIMAL PRODUCTS, AND WASTES.

GROUP 1.—*Live animals, not otherwise provided for.*

170. Stallions; geldings; mares; mules; asses; each; ten dollars.
    Provided, That sucking foals, following their dams, shall be free of duty.
171. Bovine animals:
    (a) Bulls; cows; oxen; each; two dollars.
    (b) Sucking calves, each, one dollar.
172. Swine, per head, one dollar.
173. Sucking pigs, each, twenty-five cents.
174. Animals; fish; reptiles; insects; not otherwise provided for; fifteen per centum ad valorem.
175. Birds, including poultry, each ten cents.

GROUP 2.—*Hides, skins, leather wares, intestines, and wastes.*

176. Hides and skins, raw, green, or dry, but not tanned, ten per centum ad valorem.
177. Hides and skins, tanned, with the wool or hair on; fur skins with the fur on, tanned or not; twenty-five per centum ad valorem.
178. Hides and skins, tanned, without the wool or hair, curried, dyed, or not:
    (a) Cow, and hides not otherwise provided for, split or not, of the classes known as common sole, skirting, harness, or hydraulic leather; sheep skins (basils); boot and shoe findings of any of the same; ten per centum ad valorem.
    (b) The same, of other classes; calf, goat kid, lamb, and similar skins; sheep skins finished in imitation of any of these; not having the artificial finishes enumerated under clause (c) of this paragraph; cow hide embossed in imitation of pigskin; boot and shoe findings of any of the same; fifteen per centum ad valorem.
    (c) Hides and skins, not otherwise provided for; hides and skins enameled, gilt, bronzed, bleached, figured, engraved, or embossed (except as provided in clause (b) of this paragraph); chamois, vellum, and parchment leathers; boot and shoe findings of any of the same; twenty-five per centum ad valorem.

179. Gloves:
    (a) Of kid skin, forty per centum ad valorem.
    (b) Other, twenty-five per centum ad valorem.
180. Boots and shoes:
    (a) Of cow hide, horse hide, sheep skin, and canvas, fifteen per centum ad valorem.
    (b) Other; slippers; sandals; alpargatas; of whatever material (except silk); twenty-five per centum ad valorem.
    (c) The same, of silk, fifty per centum ad valorem.
181. Saddlery and harness; parts therefor; not otherwise provided for:
    (a) Draft harness and parts therefor, twenty per centum ad valorem.
    (b) Other harness, saddlery, and harness makers' wares; parts therefor; manufactures of raw hide, not otherwise provided for; whips of whatever material; twenty-five per centum ad valorem.
182. Manufactures of leather, not otherwise provided for, thirty-five per centum ad valorem.
183. Bladders; integuments and intestines of animals; fish sounds; not otherwise provided for:
    (a) Not further advanced in manufacture than dried, thirty per centum ad valorem.
    (b) Further advanced, fifty per centum ad valorem.
184. Animal wastes and by-products not otherwise provided for:
    (a) Unmanufactured, including any of the same ground or prepared as fertilizers or as food for animals, ten per centum ad valorem.
    (b) Manufactured, or otherwise advanced in value or condition, twenty per centum ad valorem.

CLASS XII.—INSTRUMENTS, APPARATUS, MACHINERY, VEHICLES, AND BOATS.

GROUP 1.— *Musical instruments, watches, and clocks.*

185. Musical instruments; parts, appurtenances, and accessories therefor, including strings and wires; automatic devices for the production of music only; piano stools, metronomes, tuning hammers, tuning forks, pitch pipes, and similar articles for use in connection therewith; not otherwise provided for; twenty-five per centum ad valorem.
186. Instruments and machines combining other mechanical operations with the production of music, such as slot machines of that character; phonographs, gramophones, graphophones, and similar apparatus; kinetoscopes, biographs, cinematographs, magic lanterns, and similar picture-projecting devices, not otherwise provided for; parts, appurtenances, and accessories therefor; thirty-five per centum ad valorem.
187. Clocks, chronometers, watches, cyclometers, pedometers, odometers, and similar devices; cases, crystals, movements, parts, and accessories therefor; not otherwise provided for; twenty-five per centum ad valorem.

GROUP 2.—*Apparatus and machinery.*

188. Typewriters, mimeographs, Roneos, and other writing, duplicating, and manifolding machines and devices; adding machines, comtographs, and other computing apparatus; fare registers; detached parts therefor, including ribbons, pads, stencil sheets, mimeograph silks, and similar accessories therefor; stamp pads; fifteen per centum ad valorem.
189. Cash registers: detached parts therefor; twenty-five per centum ad valorem.
190. Sewing machines; detached parts therefor (except needles); fifteen per centum ad valorem.
191. Automatic slot machines, not otherwise provided for; detached parts therefor; subject to the provisions of Sec. six of this act; thirty-five per centum ad valorem.
192. Machinery and apparatus for weighing; detached parts therefor; not otherwise provided for; twenty per centum ad valorem.
193. Electric and electro-technical machinery, apparatus, and appliances:
    (a) Dynamos, generators, generating sets, alternators, motors, and similar machinery, not otherwise provided for; transformers and storage batteries; switch boards and switches; arc lamps, telephone and telegraph instruments; fans, buzzers, and annunciators; ammeters, voltmeters, wattmeters, and similar measuring apparatus; dry and wet batteries; detached parts therefor; and articles used exclusively in the installation thereof; insulators, and insulating compounds and materials used exclusively for electrical purposes; carbon; incandescent bulbs and tubes; ten per centum ad valorem.
    (b) Cooking and heating apparatus and utensils; chandeliers; desk and table lamps; flatirons; soldering and curling irons; thermo-cauteries and cauterizing instruments; surgical, dental, and therapeutic appliances, including so-called electric belts; X-ray machines; vibratory apparatus; electro-plating outfits; cigar lighters; other instruments, implements, utensils, and articles used in connection with, for, or by the application or production of electro-technical, thermo-electric, galvanic, or galvanomagnetic force; detached parts therefor; not otherwise provided for; twenty per centum ad valorem.
194. Machinery and apparatus:
    (a) Of iron, steel, or wood, for use in the crushing, handling, or conveying of sugar cane or its products in or around sugar mills; detached parts therefor; thirty per centum ad valorem.
    (b) Engines, tenders, motors, steam boilers, pumps, and machinery; diving suits; common tools, implements, and apparatus; detached parts therefor; not otherwise provided for; shafting and gearing; of iron, steel, or wood; fifteen per centum ad valorem.
    (c) The same, of other materials; emery cloth; twenty per centum ad valorem.

195. Machine belting of whatever material; ten per centum ad valorem.
196. Fine tools, implements, and instruments, of whatever material, used in the arts, trades, and professions, such as measuring instruments, micrometric gauges, mathematical and drawing instruments, manicure instruments (not pocket cutlery), watchmakers', jewelers', surgeons', dentists', engravers', carvers', glass cutting, and similar tools, instruments, and implements; detached parts therefor; not otherwise provided for, twenty per centum ad valorem.

GROUP 3.— *Vehicles*.

197. Wagons and carts for transporting merchandise; warehouse trucks; hand carts and wheelbarrows; detached parts therefor, not otherwise provided for; fifteen per centum ad valorem.
198. Automobiles.
     (a) For the transportation of merchandise, fifteen per centum ad valorem.
     (b) Other, twenty per centum ad valorem.
     (c) Detached parts and accessories for automobiles, including tires, lamps, and horns, twenty-live per centum ad valorem.
199. Bicycles, velocipedes, and motor cycles; detached parts and accessories therefor, including tires and lamps: twenty per centum ad valorem.
200. Vehicles for use on railways and tramways, and detached parts thereof:
     (a) For public or common carriers, ten per centum ad valorem.
     (b) Other, thirty per centum ad valorem.
201. Other wheeled vehicles, including perambulators; aerial machines; balloons; detached parts therefor, not otherwise provided for; twenty per centum ad valorem.
202. Detached wooden parts for any of the articles classified under paragraph one hundred and ninety-seven or paragraph two hundred and one.
     (a) Unfinished, fifteen per centum ad valorem.
     (b) Finished, twenty per centum ad valorem.

GROUP 4.—*Boats and other water craft*.

203. Boats, launches, lighters, and other water craft, set up or knocked down, imported into the Philippine Islands; cost of repairs made in foreign countries to vessels, or to parts thereof, documented for the Philippine coastwise trade or plying exclusively in Philippine waters, and for which repairs adequate facilities are afforded in the Philippine Islands, fifty per centum ad valorem.
     Provided, That upon proof, satisfactory to the collector of customs, that adequate facilities are not afforded in the Philippine Islands for such repairs, the same shall be subject to the provisions of paragraph three hundred and fifty-one of this act; and

Provided further, That if the owner or master of such vessel shall furnish evidence, satisfactory to the collector of customs, that such vessel, while in the regular course of her voyage, was compelled by stress of weather or other casualty to put into a foreign port or place and make such repairs to secure the safety of the vessel or to enable her to return to the Philippine Islands, such duty shall not be imposed; and

Provided further, That furnishings, stores, and supplies not otherwise provided for, purchased abroad and imported in such vessels, shall be dutiable under the corresponding paragraphs of this act.

NOTE.—The expression "imported into the Philippine Islands" shall be held to mean "brought into the jurisdictional waters of the Philippine Islands in or on another vessel, or towed therein by another vessel (except when becalmed or disabled at sea), as distinguished from coming into said islands under the craft's own steam, sail, or other motive power."

CLASS XIII.—ALIMENTARY SUBSTANCES.

GROUP I.—*Poultry, meats, soups, and fish.*

204. Poultry; game; not otherwise provided for; dressed or not; gross weight, one hundred kilos, four dollars.
205. Meat, fresh, not otherwise provided for, gross weight, one hundred kilos, one dollar.
206. Meat; sausage casings; salted or in brine; gross weight, one hundred kilos, two dollars and fifty cents.
207. Hams; bacon; other meats and sausages; dry, cured, or smoked; not preserved in cans; including weight of immediate containers, one hundred kilos, four dollars and fifty cents.

   Provided, That sausages classified under this paragraph may be imported in any kind of package exceeding in weight ten kilos each; and

   Provided further, That salt used for the packing of any article classified under this paragraph shall be dutiable under clause (c) of paragraph seventy-four.
208. Lard; imitations thereof; gross weight, one hundred kilos, two dollars and fifty cents.
209. Canned or potted meats, such as beef, veal, mutton, lamb, pork, ham, and bacon, plainly prepared and simply preserved, not otherwise provided for; common preparations thereof, with or without vegetables or other simple ingredients, such as Irish stew, corned beef hash, Chili-con-carne, hog and hominy, dry chipped beef, and the like, fifteen per centum ad valorem.
210. Internal parts of animals, such as tongue, liver, and tripe; rabbits; poultry; ordinary preparations thereof; canned or potted; sausages, not otherwise provided for, twenty per centum ad valorem.
211. Canned or potted game; paté de foie gras; deviled ham, meats or game, and preparations thereof; mince-meat; meat patés; jellied lambs' and sheeps' tongues; boneless pigs' feet; sweet

breads; brains; similar products of delicatessen class; not otherwise provided for, twenty-five per centum ad valorem.

212. Canned or potted soups and broths; clam chowder; fifteen per centum ad valorem.

213. Meat extracts in any form; meat juice and soup tablets; condensed or concentrated soup preparations, dry or in paste, twenty-five per centum ad valorem.

214. Salted or dried codfish, gross weight, one hundred kilos, one dollar and sixty cents.

215. Fish, in cans, glass, or jars:
    (a) Cod, herring, mullet, haddock, salmon, and mackerel, plainly prepared and simply preserved; sardines in oil or tomato sauce, fifteen per centum ad valorem.
        otherwise provided for, twenty per centum ad valorem.
    (c) Fish, shell fish, sea food, and preparations thereof, such as anchovies, merluza, angulas, awabi, sardines not otherwise provided for, lampreys, whiting, turtle, fish roe, eels in jelly, sharks' fins in any form, shrimp, bloater, and fish pastes and butters; similar products of delicatessen class, twenty-five per centum ad valorem.

216. Fish, not otherwise provided for:
    (a) Fresh, with only the salt indispensible for preservation, gross weight, one hundred kilos, two dollars and ninety cents.
    (b) Dried, salted, smoked, or pickled, in bulk; gross weight, one hundred kilos, two dollars and twenty-five cents.

217. Oysters, clams, and shell fish, in bulk, not otherwise provided for; fresh oysters in cans; gross weight; one hundred kilos, five dollars.

GROUP 2.—*Grains, seeds, forage, cereals, and preparations thereof.*

218. Rice, gross weight (until May first, nineteen hundred and ten):
    (a) Unhusked, one hundred kilos, sixty cents.
    (b) Husked, one hundred kilos, one dollar.
    (c) Flour, one hundred kilos, two dollars.
    On and after May first, nineteen hundred and ten:
    (a) Unhusked, one hundred kilos, eighty cents.
    (b) Husked, one hundred kilos, one dollar and twenty cents.
    (c) Flour, one hundred kilos, two dollars.
    Provided, That the Governor-General, by and with the advice and consent of the Philippine Commission, may, in his discretion, continue in force the rates of duty first prescribed in this paragraph until, in his judgment, conditions in the Philippine Islands may warrant the imposition of the higher rates herein prescribed; and
    Provided further, That the Governor-General, by and with the advice and consent of the Philippine Commission, may suspend all duties upon rice, or the duties upon rice for consumption in particular localities, to be designated by him, whenever and for such period as, in his judgment, local conditions require; in which event, rice admitted free by virtue of his order shall be distributed under governmental supervision, or in accordance with such regulations as he may prescribe.

219. Wheat, rye, and barley, gross weight:
    (a) In grain, one hundred kilos, twenty-five cents.
    (b) In flour, one hundred kilos, forty-seven cents.
220. Corn (maize), oats, and millet; cereals and grains not otherwise provided for; gross weight:
    (a) In grain, one hundred kilos, seventeen cents.
    (b) In meal or flour, not otherwise provided for, one hundred kilos, eighty-three cents.
221. Cereals prepared for table use, such as oatmeal, cornmeal, cracked wheat, cornstarch, and similar preparations, not otherwise provided for, ten per centum ad valorem.
222. Malted milk, infants' foods; similar preparations; fifteen per centum ad valorem.
223. Bread, biscuits, crackers and wafers, of flour of cereals or pulse; including weight of immediate containers:
    (a) Unsweetened, one hundred kilos, three dollars.
    (b) Sweetened, one hundred kilos, five dollars.
224. Cakes and puddings, twenty-five per centum ad valorem.
225. Vermicelli; macaroni; pastes for soup, not otherwise provided for; including weight of immediate containers; one hundred kilos, two dollars and fifty cents.
226. Birds' nests, edible, thirty per centum ad valorem.
227. Seeds, not otherwise provided for, gross weight, one hundred kilos, one dollar.
228. Hay; bran; forage; straw, not otherwise provided for; seeds and unhusked grains, cracked, or otherwise prepared for animal food; oil cake; five per centum ad valorem.

GROUP 3.— *Pulse, vegetables, fruits, and nuts.*

229. Dried beans, pease, and other pulse:
    (a) In bulk, gross weight, one hundred kilos, eighty cents.
    (b) In small or retail packages, including weight of immediate containers, one hundred kilos, two dollars and sixty-five cents.
    (c) In flour, gross weight, one hundred kilos, one dollar and fifty cents.
230. Vegetables fresh (except onions and Irish potatoes), gross weight, one hundred kilos, one dollar.
231. Vegetables, dried or dessicated, not otherwise provided for:
    (a) In bulk, gross weight, one hundred kilos, one dollar and thirty cents.
    (b) In small or retail packages, including weight of immediate containers, one hundred kilos, two dollars and twenty-five cents.
232. Vegetables, preserved, not otherwise provided for:
    (a) In bulk, gross weight, one hundred kilos, one dollar.
    (b) In small or retail packages, including weight of immediate containers; one hundred kilos, one dollar and fifty cents.
        Provided, That no article classified under clause (b) of this paragraph, shall pay a less rate of duty than fifteen per centum ad valorem.
233. Vegetables, pickled:
    (a) In bulk, gross weight, one hundred kilos, one dollar and fifty cents.

(b) In small or retail packages, including weight of immediate containers, kilo, three cents.

> *Provided,* That no article classified under clause (b) of this paragraph shall pay a less rate of duty than fifteen per centum ad valorem.

234. Fruits, fresh, gross weight, one hundred kilos, one dollar and twenty-five cents.

235. Fruits, dried:

(a) In bulk, gross weight, one hundred kilos, one dollar and fifty cents.

(b) In small or retail packages, including weight of immediate containers, one hundred kilos, two dollars and fifty cents.

> *Provided,* That no article classified under clause (b) of this paragraph shall pay a less rate of duty than fifteen per centum ad valorem.

236. Fruits, preserved, not otherwise provided for:

(a) In bulk, gross weight, one hundred kilos, one dollar and fifty cents.

(b) In small or retail packages, including weight of immediate containers, one hundred kilos, two dollars.

> *Provided,* That no article classified under clause (b) of this paragraph shall pay a less rate of duty than fifteen per centum ad valorem.

237. Fruits, in jellies, jams, marmalades, butters, and similar preparations; fruit pulp; twenty per centum ad valorem.

238. Fruits, brandied, or similarly preserved; fruits conserved or crystallized; twenty-five per centum ad valorem.

239. Nuts; nut products; not otherwise provided for: twenty-five per centum ad valorem.

GROUP 4.—*Sugar, molasses, glucose, and confectionery.*

240. Sugar:

(a) Raw, gross weight, one hundred kilos, three dollars and seventy cents.

(b) Refined, including weight of immediate containers, one hundred kilos, four dollars and twenty cents.

241. Molasses and sirups, not otherwise provided for; honey:

(a) In bulk, gross weight, one hundred kilos, two dollars.

(b) In small or retail packages, including weight of immediate containers, one hundred kilos, three dollars.

242. Glucose, gross weight, one hundred kilos, one dollar and sixty cents.

243. Saccharine, including weight of immediate containers, kilo, two dollars.

244. Candies; confectionery; sweetmeats; chewing gum; not otherwise provided for: twenty-five per centum ad valorem.

GROUP 5.—*Coffee, tea, cacao, spices, sauces, condiments, and flavoring extracts.*

245 Coffee:

(a) Unroasted, gross weight, one hundred kilos, five dollars and thirty cents.

(b) Roasted, ground or not, gross weight, one hundred kilos, seven dollars.

(c) In packages weighing each less than three kilos, including weight of immediate containers, one hundred kilos, nine dollars.

246. Chicory, gross weight, one hundred kilos, four dollars and twenty cents.

247. Tea, including weight of immediate containers, kilo, fifteen cents.

248. Cacao:
  (a) Unground, gross weight, one hundred kilos, seven dollars and twenty cents.
  (b) Other; cacao butter; including weight of immediate containers; one hundred kilos, twelve dollars and fifty cents.
    Provided, That no article classified under clause (b) of this paragraph shall pay a less rate of duty than twenty-five per centum ad valorem.

249. Chocolate, including weight of immediate containers:
  (a) In forms or lumps for manufacturing purposes, one hundred kilos, ten dollars.
  (b) In cakes or powder, kilo, fifteen cents.
    Provided, That no article classified under clause (b) of this paragraph shall pay a less rate of duty than twenty-five per centum ad valorem.

250. Cinnamon, cloves, allspice, and mace, including weight of immediate containers:
  (a) Underground, one hundred kilos, eight dollars.
  (b) Ground, one hundred kilos, ten dollars.

251. Nutmegs, including weight of immediate containers:
  (a) Unhusked, kilo, three cents.
  (b) Husked, kilo, five cents.
  (c) Ground, kilo, eight cents.

252. Pepper, white or black; pod peppers, dried; including weight of immediate containers:
  (a) Whole, one hundred kilos, two dollars and twenty cents.
  (b) Ground, kilo, eight cents.

253. Mustard; horse-radish; including weight of immediate containers:
  (a) Unground, kilo, two cents.
  (b) Ground, kilo, six cents.
  (c) In paste, kilo, ten cents.

254. Saffron, including weight of immediate containers, kilo, four dollars.

255. Spices, not otherwise provided for, including weight of immediate containers:
  (a) Unground, one hundred kilos, eight dollars.
  (b) Ground; curry powder; one hundred kilos, ten dollars.
    Provided, That no article classified under this paragraph shall pay a less rate of duty than twenty-five per centum ad valorem.

256. Sauces for table use, not otherwise provided for, such as tomato, caper, tobasco, Worcestershire, catsup, and like preparations, twenty-five per centum ad valorem.

257. Vinegar:
  (a) In receptacles containing each more than two liters, liter, two cents.
  (b) In other receptacles, liter, three cents.

258. Flavoring extracts, compounds, and sirups; including weight of
immediate containers:
   (a) Without alcohol or containing not to exceed fourteen per
centum of alcohol, kilo, twenty-five cents.
   (b) Containing more than fourteen per centum of alcohol, kilo,
thirty-five cents.
      Provided, That any article, otherwise subject to classifica-
tion under this clause, containing more than twenty-four per
centum of alcohol shall be classified under paragraph two
hundred and sixty-four.
259. Vanilla beans, including weight of immediate containers, two
dollars and fifty cents.

GROUP 6.—*Spirits, wines, malt, and other beverages.*

NOTE.—For the purpose of assessment under those paragraphs in which the proof liter is the basis, each and every
gauge or wine liter of measurement shall be counted as at
least one proof liter. All imitations of whisky, rum, gin,
brandy, spirits, or wines, imported by or under any names
whatsoever shall be subjected to the highest rate of duty
provided for the genuine articles respectively intended to
be represented, with a surtax of fifty per centum.
260. Alcohol, liter, twenty-five cents.
261. Whisky; rum; gin; brandy; other spirits; not otherwise provided
for; proof liter, fifty cents.
262. Blackberry and ginger brandy, proof liter, thirty cents.
263. Cocktails, liqueurs, cordials, and other compounded spiritous
beverages and bitters, not otherwise provided for, proof
liter, sixty-five cents.
264. Wines, sparkling, liter, one dollar.
265. Still wines; sake, containing fourteen per centum or less of
alcohol:
   (a) In receptacles containing each more than two liters, liter,
two cents.
   (b) In receptacles containing each two liters or less, liter, seven
and one-half cents.
      Provided, That no article classified under this paragraph
shall pay a less rate of duty than forty per centum ad valorem.
266. Still wines; sake; containing more than fourteen per centum of
alcohol:
   (a) In receptacles containing each more than two liters, liter,
fifteen cents.
   (b) In receptacles containing each two liters or less, liter,
twenty-five cents.
      Provided, That no article classified under this paragraph
shall pay a less rate of duty than fifty per centum ad valorem; and
      Provided further, That any of such articles containing
more than twenty-four per centum of alcohol shall be classified under paragraph two hundred and sixty-four.
267. Malt beverages; ciders:
   (a) In receptacles containing each more than two liters, hectol.,
two dollars and twenty-five cents.
   (b) In other receptacles, hectol., four dollars and ninety cents.

268. Sweetened, flavored, or aerated waters; natural mineral waters, aerated or not; ginger ale; root beer; unfermented fruit juice; non-alcoholic beverages, not otherwise provided for; hectol., one dollar and fifty cents.

269. Fruit juice, pure or with sufficient sugar to preserve it, without alcohol or containing not more than four per centum of alcohol, liter, five cents.

GROUP 7.—*Various.*

270. Milks and creams, pure, or with sufficient sugar to preserve them, ten per centum ad valorem.

271. Milks and creams, compounded with other substances; milk powders and tablets; not otherwise provided for; twenty per centum ad valorem.

272. Eggs, not otherwise provided for:
    (a) Fresh or preserved, in natural form, gross weight, one hundred kilos, one dollar.
    (b) Egg powders; other preparations of eggs; not otherwise provided for; twenty-five per centum ad valorem.

273. Cheese of all kinds and substitutes therefor, fifteen per centum ad valorem.

274. Butter, including weight of immediate containers, kilo, six cents.

275. Oleomargarine; butterine; ghee; imitations of butter; including weight of immediate containers, kilo, eight cents.

276. Articles and products edible by mankind, not otherwise provided for:
    (a) Crude and in natural state, ten per centum ad valorem.
    (b) Prepared, preserved, or advanced in value or condition by any process or manufacture, twenty per centum ad valorem.

CLASS XIV.—MISCELLANEOUS.

277. Fans, of all kinds, thirty-five per centum ad valorem.

278. Pens, not otherwise provided for; needles (except surgical needles); common and safety pins; hooks and eyes; button rings and fasteners; crochet hooks; hairpins; of common metals (except those covered or coated with gold or silver); twenty-five per centum ad valorem.

279. Trinkets and ornaments of all kinds (except those of gold or silver, or of gold or silver plate, or in which the component material of chief value is amber, jet, jade, tortoise shell, coral, ivory, meerschaum, or mother-of-pearl), including weight of immediate containers, kilo, one dollar and twenty-five cents: *Provided,* That no article classified under this paragraph shall pay a less rate of duty than thirty per centum ad valorem.

280. Amber; jet; tortoise shell; coral; ivory; meerschaum; mother-of-pearl:
    (a) Unwrought; cut for settings or pierced for beads; fifteen per centum ad valorem.
    (b) Wrought, not otherwise provided for, thirty-five per centum ad valorem.

281. Horn; bone; whalebone; celluloid; imitations thereof, or of any
of the substances enumerated in paragraph two hundred and
eighty-one; including weight of immediate containers:
> (a) Unwrought, kilo, thirty cents.
> (b) Wrought, not otherwise provided for, kilo, one dollar and
> twenty-five cents.
>> *Provided*, That no article classified under clause (b) of
>> this paragraph shall pay a less rate of duty than thirty per
>> centum ad valorem.

282. Artificial teeth, with plates or not; artificial eyes; artificial limbs
and members; similar articles for the alleviation of the
inconveniences resulting from physical defects; ten per
centum ad valorem.

283. Buttons, including weight of immediate containers:
> (a) Of mother-of-pearl, kilo, one dollar and thirty cents.
> (b) Of bone, porcelain, composition, wood, steel, iron, or simi-
> lar materials, kilo, thirty cents.
> (c) Of other materials (except gold, silver or platinum, or
> gold or silver plate), kilo, fifty cents.
>> *Provided*, That no article classified under this paragraph
>> shall pay a less rate of duty than twenty-five per centum
>> ad valorem.

284. Shells, not otherwise provided for:
> (a) Not further advanced in condition than polished, ten per
> centum ad valorem.
> (b) Further advanced; manufactures in which shells, not other-
> wise provided for, are the component material of chief
> value; twenty-five per centum ad valorem.

285. Sponges, natural, including hexactinellida and loofah:
> (a) Not further advanced in condition than washed or bleached,
> twenty-five per centum ad valorem.
> (b) Further advanced; manufactures in which sponge or loofah
> is the component material of chief value; forty per cen-
> tum ad valorem.

286. Felt or textiles prepared or coated with tar, pitch, or similar
substances; rubberoid; similar materials; for roofing, sheathing, and structural purposes: gross weight, one hundred kilos,
ninety cents.

287. Oilcloth (except of silk); linoleum; corticine:
> (a) In the piece, fifteen per centum ad valorem.
> (b) Made up into articles, twenty-five per centum ad valorem.

288. Tool bags, chests, and cases; trunks; valises; suit cases; traveling bags; telescopes; hat boxes; and similar receptacles for
personal effects; shawl straps; of whatever material: twenty-
five per centum ad valorem.

289. Stuffed or mounted birds or animals, not otherwise provided for,
twenty per centum ad valorem.

290. Feathers for ornaments; stuffed birds or animals or parts thereof
for use on wearing apparel or for toilet purposes, natural, finished, or manufactured, sixty per centum ad valorem.

291. Feathers and downs, not otherwise provided for:
> (a) Not further advanced in condition than cleaned, twenty
> per centum ad valorem.

(b) Further advanced; manufactures in which feathers or downs are the component material of chief value; forty per centum ad valorem.

292. Artificial flowers, buds, pistils, leaves, fruits, seeds, and moss; other parts of artificial fruits and flowers; of whatever material; fifty per centum ad valorem.

293. Caoutchouc and gutta-percha:

(a) Crude; rubber, in sheets, sheeting, or packing, even with cloth or wire insertions; gaskets and washers; ten per centum ad valorem.

(b) Rubber, soft, in articles not otherwise provided for, twenty-five per centum ad valorem.

(c) Rubber, hard, in articles not otherwise provided for, thirty per centum ad valorem.

294. Hose and flexible tubing, of whatever dimensions or materials, fifteen per centum ad valorem.

295. Reservoir pens; parts and points therefor; of whatever material; twenty-five per centum ad valorem.

296. Games and toys, including face masks, paper hats and canes, artificial Christmas trees, Christmas-tree decorations, toy carts and other small vehicles for children's use not otherwise provided for; diminutive articles for use as toys, not adapted for practical purposes; including weight of immediate containers, kilo, ten cents.

Provided, That no article of gold, silver, or platinum, or of gold or silver plate, or of tortoise shell, coral, ivory, or mother-of-pearl, shall be classified under this paragraph; and

Provided further, That no article classified under this paragraph shall pay a less rate of duty than twenty-five per centum ad valorem.

297. Golf sticks; polo mallets; tennis rackets; baseball and cricket bats; balls of all kinds for use in the sports (except bowling, billiard, pool, and bagatelle balls); fencing masks and foils; gymnasium apparatus; croquet sets, and parts therefor; twenty-five per centum ad valorem.

298. Umbrellas and parasols:

(a) Covered with paper, each, eight cents.

(b) Covered with silk, each, fifty cents.

(c) Covered with other stuffs, each, twenty cents.

(d) Umbrella frames, uncovered, mounted on sticks, forty per centum ad valorem.

Provided, That no article classified under this paragraph shall pay a less rate of duty than twenty-five per centum ad valorem.

299. Hats; bonnets; crowns therefor; of straw, chip, palm leaf, grass, rattan, osiers, and analogous materials:

(a) Complete, not trimmed, each, thirteen cents.

(b) The same, trimmed, each, twenty-two cents.

(c) Crowns for, each, eleven cents.

Provided, That no article classified under this paragraph shall pay a less rate of duty than twenty-five per centum ad valorem.

300. The same, of other materials:
    (a) Complete, not trimmed, each, twelve cents.
    (b) The same, trimmed, each, twenty cents.
    (c) Crowns for, each, eleven cents.
      Provided, That no article classified under this paragraph shall pay a less rate of duty than twenty-five per centum ad valorem.

301. Caps; fezzes; turbans; headgear not otherwise provided for; thirty per centum ad valorem.

302. Cameras and parts therefor; photographic equipment and articles for use in photography, not otherwise provided for, such as lenses, tripods, photographic plates and films, film packs and kits, plate holders and frames, developing lights, baths, and trays; twenty per centum ad valorem.

303. Appliances and apparatus, parts and cases therefor and accessories thereto, not otherwise provided for, for mathematical, optical, astronomical, surgical, geodetical, and other scientific purposes, such as thermometers, barometers, alcoholometers, salmometers, hydrometers, vacuometers, radiometers, appliances for sight testing, microtomes, telescopes, microscopes and their slide glasses, stethoscopes, theodolites, transits, sextants, quadrants, compasses, and the like, twenty-five per centum ad valorem.

304. Tobacco:
    (a) Wrapper tobacco; filler tobacco, when mixed or packed with more than fifteen per centum of wrapper tobacco; leaf tobacco, the product of two or more countries or dependencies, mixed or packed together; unstemmed, kilo, four dollars and eighty cents.
    (b) The same, stemmed, kilo, five dollars and fifty cents.
    (c) Filler tobacco, not otherwise provided for, unstemmed, kilo, seventy-seven cents.
    (d) The same, stemmed, kilo, one dollar and two cents.
    (e) Cigars, cigarettes, cheroots; paper cigars and cigarettes, including wrappers; kilo, nine dollars and ninety cents and twenty-five per centum ad valorem.
    (f) Other tobacco, manufactured or unmanufactured, not otherwise provided for, kilo, one dollar and twenty cents.
      NOTE.—The term "wrapper tobacco," as used in this act, shall be taken to mean that quality of leaf tobacco which is suitable for cigar wrappers, and the term "filler tobacco" shall be taken to mean all other leaf tobacco.

305. Wastes, not otherwise provided for, ten per centum ad valorem.

306. Materials, substances, and articles, not otherwise provided for:
    (a) Not advanced in value or condition by any process or manufacture, ten per centum ad valorem.
    (b) Further advanced, but not manufactured into articles, fifteen per centum ad valorem.
    (c) Manufactured into articles, twenty-five per centum ad valorem.

307. Cost of repairs upon articles of easy identification (except those provided for in paragraph two hundred and three), exported from the Philippine Islands and reimported therein, twenty-five per centum ad valorem.

Provided, That any such article, exclusive of the repairs thereon, shall be free of duty when reimported, upon the compliance with the regulations of the Insular collector of customs governing such exportations and reimportations; otherwise, the terms of section twelve shall apply.

FREE LIST.

SEC. 12. That the following articles shall be free of duty upon importation thereof into the Philippine Islands:

308. Trees; shoots; plants; moss; live.
309. Ores, and scoriae resulting from the smelting thereof; filings, cuttings, and other wastes, of common metals, resulting from manufacture, and fit only for resmelting; scrap iron, copper, brass, tin, zinc, and lead, and combinations thereof; bell metal; copper regulus; copper matte; cast or malleable iron in pigs; soft or wrought iron in ingots; steel in ingots; tin, lead, zinc, nickel, and aluminum, in pigs, lumps, or ingots; Muntz metal.
310. Articles, such as anchors, binnacles, propellers, and the like, the character of which, as imported, prevents their use for other than the construction, equipment, or repair of vessels; life preservers and life buoys.
311. Oakum.
312. Raw cotton.
313. Vegetable fibers, raw or hackled, not otherwise provided for.
314. Bristles; animal hair and wool; not further advanced in condition than washed.
315. Paper pulp and paper stock.
316. Samples of the kind, in such quantity, and of such dimensions or construction as to render them unsalable or of no appreciable commercial value; models not adapted for practical use.
317. Onions; Irish potatoes; in natural state.
318. Gold; silver; platinum; in bars, sheets, pieces, dust, scrap, or in broken-up jewelry or table service.
319. Hops and malt.
320. Coins and currency of national issue; executed checks, drafts, bills of exchange and similar commercial documents.
321. Natural manures.
322. Cinchona bark; sulphate and bisulphate of quinine; alkaloids and salts of cinchona bark; in whatever form.
323. Telegraph cables, of the class known as submarine.
324. Vaccines and serums.
325. Ice.
326. Hand paintings in oil, water color, or pastel; pen and ink drawings; for use as works of art, and not as a decoration of merchandise, nor for use in manufacture or the industrial arts and sciences; photographs, paintings, crayons, and other pictorial representations of actual persons, either living or deceased.
327. Lithographs, posters, calendars, and signs, framed (when the frame bears sufficient advertising matter to render it of no commercial value) or not; pamphlets, booklets, and folders;

for advertising purposes only, and having no commercial value.

NOTE.—Store, office, and business signs, *i. e.*, those for advertising local business houses, firms, offices, associations, corporations, trades or professions, shall not be classified under this paragraph.

328. Magazines, reviews, newspapers, and like published periodicals; bibles and extracts therefrom; hymnals and hymns for religious uses; books and music in raised print, used exclusively by the blind; text-books prescribed for use in any school in the Philippine Islands: *Provided*, That complete books published in parts in periodical form shall not be classified under this paragraph.

329. Public documents issued by foreign governments; correspondence, manuscripts and typewritten documents; not prohibited by section six of this act; collections of stamps of national issue, used or unused.

330. Medals, badges, cups, and other small articles actually bestowed as trophies or prizes, or those received or accepted as honorary distinctions.

331. Pipe organs imported for the bona fide use of and by the order of any society incorporated or established for religious or educational purposes, or expressly for presentation thereto; *Provided*, That the terms of this paragraph shall be retroactive and of full force and effect from and after January 1, 1909, anything in this act to the contrary notwithstanding; and *provided further*, That any duty paid upon any pipe organ so imported, since said date, shall be subject to refund.

FREE SUBJECT TO EXPRESS CONDITIONS.

SEC. 13. That the following articles shall be free of duty upon the importation thereof into the Philippine Islands upon compliance with the formalities prescribed in each paragraph:

332. Eggs and cocoons of the silk worm; subject to exclusion if diseased or for other cause.

333. Breeding animals of a recognized breed, duly registered in the book of record established for that breed: *Provided*, That certificate of such record, and pedigree of such animal, duly authenticated by the proper custodian of such book or record, shall be produced and submitted to the collector of customs, together with affidavit of the owner or importer, that such animal is the identical animal described in said certificate of record and pedigree.

334. Carabao, and other bovine work animals; mules; until such time as the governor-general shall certify that conditions in the Philippine Islands warrant the imposition of duty thereon in accordance with the rates prescribed in group one of Class XI of this act.

335. Commercial samples, the value of any single importation of which does not exceed five thousand dollars, upon the filing of a bond in an amount equal to double the ascertained duties thereon, with sureties satisfactory to the collector of customs,

conditioned for the exportation of said samples within six months from the date of their importation, or in default thereof, the payment of the corresponding duties thereon. If the value of any single consignment of such commercial samples exceeds five thousand dollars, the importer thereof may select any portion of same not exceeding in value five thousand dollars, for entry under the provisions of this paragraph, and the remainder of the consignment may be entered in bond, or for consumption as the importer shall elect.

336. Regalia; gems; statuary; specimens or casts of sculptures; imported for the bona fide use of and by the order of any society incorporated or established solely for religious, philosophical, educational, scientific, or literary purposes, or for the encouragement of the fine arts, or for the use of and by the order of any college, academy, school, or seminary of learning, or of any public library; not for barter, sale, or hire: *Provided*, That the term "regalia" shall be held to include only such insignia of rank or office, or emblems, as may be worn upon the person or borne in the hand during public exercises of the society or institution, and shall not include articles of furniture, fixtures, or ordinary wearing apparel, nor personal property of individuals.

337. Works of art, including pictorial paintings on glass (except stained windows or window glass), imported expressly for presentation to a governmental institution, or to any municipal or provincial corporation, or to any incorporated or established religious society, college, or other public institution.

338. Supplies and materials imported by, or for the use of, the United States Government, or the government of the Philippine Islands, or any of their subordinate branches: *Provided*, That title shall have passed to one of said governments before such supplies are released from customs custody, and *Provided further*, That any article (except those enumerated in paragraph two hundred and eighteen), which would otherwise be classified under this paragraph, shall, if imported for sale to private or corporate persons, be dutiable under the corresponding paragraph of this act.

339. Wearing apparel, articles of personal adornment, toilet articles, books, portable tools and instruments, theatrical costumes, and similar personal effects, accompanying travelers or tourists, in their baggage, or arriving within a reasonable time, in the discretion of the collector of customs, before or after the owners, in use of, and necessary and appropriate for the wear or use of such persons according to their profession or position, for the immediate purposes of their journey and their present comfort and convenience: *Provided*, That this exemption shall not be held to apply to merchandise or articles intended for other persons or for barter or sale; and *Provided further*, That the collector of customs may, in his discretion, require a bond for the exportation of, or the payment of duties upon, articles classified under this paragraph, within the time and in the manner prescribed by paragraph three hundred and forty.

340. Vehicles, horses, harness, bed and table linen, table service, furniture, musical instruments, and personal effects of like character, owned and imported by travelers or tourists for their convenience and comfort; upon identification and the giving of a bond with sureties satisfactory to the collector of customs in an amount equal to double the estimated duties thereon, conditioned for the exportation thereof, or payment of the corresponding duties thereon, within four months from the date of entry: *Provided*, That the collector of customs may extend the time for exportation or payment of duties for a term not exceeding three months from the expiration of the original period.

341. Professional instruments and implements; tools of trade, occupation, or employment; wearing apparel; domestic animals; personal and household effects, including those of the kind and class provided for under paragraphs three hundred and thirty-nine and three hundred and forty, belonging to persons coming to settle in the Phillppine Islands, in quantities and of the class suitable to the profession, rank, or position of the person importing them, for their own use and not for barter or sale, accompanying such persons, or arriving within a reasonable time, in the discretion of the collector of customs, before or after the arrival of their owners; upon the production of evidence satisfactory to the collector of customs that such persons are actually coming to settle in the Philippine Islands, that the articles are brought from their former place of abode, that change of residence is bona fide, and that the privilege of free entry under this paragraph has never been previously granted to them:

Provided, That neither merchandise of any kind, nor machinery or other articles for use in manufacture shall be classified under this paragraph; and

Provided further, That officers and employees of the United States Government or of the government of the Philippine Islands, or religious missionaries taking station in the islands, shall be considered as "coming to settle" for the purposes of this paragraph.

342. Vehicles; animals, birds, insects, and fish; portable theaters; circus and theatrical equipment, including sceneries, properties, and apparel; devices for projecting pictures and parts and appurtenances therefor; panoramas, wax figures, and similar objects; for public entertainment; upon identification and the giving of a bond with sureties satisfactory to the collector of customs in an amount equal to double the estimated duties thereon, conditioned for the exportation thereof, or payment of the corresponding duties thereon within the time and in the manner prescribed by paragraph three hundred and forty.

343. Personal effects, not merchandise, of residents of the Philippine Islands dying in foreign countries; upon identification as such, satisfactory to the collector of customs.

344. Works of fine art for public museums and galleries, or for art schools; models, archaeological and numismatic objects, specimens and collections of mineralogy, botany, zoology,

and ethnology, including skeletons, fossils, and other ana-
tomical specimens for schools, academies, public museums,
and corporations and societies organized for scientific or
artistic purposes; on proof, satisfactory to the collector of
customs, of their destination.

345. Official consular supplies, consigned by a foreign government of
which the consignee is the consular representative in the
Philippine Islands, to him as such official, in an amount and
of the kind and class allowed free entry by said foreign government when consigned by the Government of the United
States of America to its consular representative within the
jurisdiction of such foreign government.

346. Pumps for the salvage of vessels; upon identification and the
giving of a bond with sureties satisfactory to the collector
of customs, in an amount equal to double the estimated
duties thereon, conditioned for the exportation thereof or
payment of the corresponding duties thereon within the
time and in the manner prescribed by paragraph three hundred and forty.

FREE UPON COMPLIANCE WITH CORRESPONDING REGULATIONS.

SEC. 14. That the following articles shall be free of duty upon the
importation thereof into the Philippine Islands upon compliance with
regulations which shall be prescribed in accord with the provisions of
each paragraph:

347. Wearing apparel and household effects, including those articles
provided for under paragraphs three hundred and thirty-nine
and three hundred and forty, belonging to residents of the
Philippine Islands, returning from abroad, which were
exported from the said islands by such returning residents
upon their departure therefrom, or during their absence
abroad, upon the identity of such articles being established
to the satisfaction of the collector of customs, under such
regulations as the insular collector of customs shall prescribe; articles of the same kind and class purchased in
foreign countries by natives of the Philippine Islands during
their absence abroad, and accompanying them upon their
return to said islands, or arriving within a reasonable time,
in the discretion of the collector of customs, before or after
their return, upon proof satisfactory to the collector of customs that the same have been in their use abroad for more
than one year.

348. Foreign articles, goods, wares, or merchandise destined for dis-
play in public expositions in the Philippine Islands; animals
for exhibition, or competition for prizes, together with the
harness, vehicles, and tackle necessary for the purposes
designated; subject to such rules, regulations, and conditions as shall be prescribed by the insular collector of customs, with respect to bonding for exportation thereof or
payment of duty thereon.

349. Philosophical, historical, economic, and scientific books; apparatus, utensils, and instruments, specially imported for the
bona fide use of and by the order of any society or institu-

tion incorporated or established solely for philosophical, educational, scientific, charitable, or literary purposes, or for the encouragement of the fine arts, or for the bona fide use of and by the order of any college, academy, school, or seminary of learning in the Philippine Islands, or of any public library, and not for barter, sale, or hire; subject to such regulations as shall be prescribed by the Insular collector of customs.

The provisions of this paragraph in respect to books shall apply to any individual importing not exceeding two copies of any one work for his own use and not for barter, sale, or hire.

350. Articles of the growth, produce, or manufacture of the Philippine Islands; paintings which are works of art; books; exported to a foreign country and returned without having been advanced in value or improved in condition by any process of manufacture or other means, and upon which no drawback or bounty has been allowed; articles returned from foreign expositions; subject to identification under such rules and regulations as the Insular collector of customs shall prescribe.

351. Repairs to vessels documented in the Philippine Islands or regularly plying in Philippine waters, made in foreign countries, upon proof, satisfactory to the collector of customs, that adequate facilities for such repairs are not afforded in the Philippine Islands.

352. Articles and materials actually used in the construction, equipment, or repair, within the Philippine Islands, of vessels, their machinery, tackle, or apparel; subject to such restrictions, conditions, and regulations as the Insular collector of customs shall prescribe.

353. Articles brought into the Philippine Islands for the purpose of having repairs made thereto, upon the filing of a bond with sureties satisfactory to the collector of customs, in an amount equal to double the estimated duties thereon, conditioned for the exportation thereof or payment of the corresponding duties thereon within a period of not to exceed six months from the date of importation thereof, in the discretion of the collector of customs; subject to such rules and regulations as the Insular collector of customs shall prescribe.

354. Coverings and holdings of articles, goods, wares, and merchandise (usual), except as expressly provided. (See Rule thirteen.)

EXPORT DUTIES.

SEC. 15. That upon the exportation from the Philippine Islands of the following articles, there shall be levied and collected thereon the following export duties:

355. Abaca (hemp), gross weight, one hundred kilos, seventy-five cents.

356. Sugar, gross weight, one hundred kilos, five cents.

357. Copra, gross weight, one hundred kilos, ten cents.

358. Tobacco, manufactured, gross weight, one hundred kilos, one dollar and fifty cents.

359. Stems, clippings, and other wastes of tobacco, gross weight, one hundred kilos, fifty cents.
360. Tobacco, raw, grown in the provinces of Cagayan, Isabela, and Nueva Vizcaya (island of Luzon), gross weight, one hundred kilos, one dollar and fifty cents.
361. Tobacco, raw, grown in other provinces of the Philippine Islands, gross weight, one hundred kilos, one dollar.

NOTE.—Certificates of origin of raw tobacco may be required by the collector of customs at the port of exportation when proof of the place of production is necessary.
362. Shells, gross weight:
(a) Tortoise; mother-of-pearl; one hundred kilos, one dollar.
(b) Other, one hundred kilos, fifty cents.

WHARFAGE.

SEC. 16. That there shall be levied and collected upon all articles, goods, wares, or merchandise, exported through ports of entry of the Philippine Islands, a duty of one dollar per gross ton of one thousand kilos, as a charge for wharfage, whatever be the port of destination or nationality of the exporting vessel; *Provided*, That articles, goods, wares, or merchandise imported, exported, or shipped in transit for the use of the Government of the United States, or of that of the Philippine Islands, shall be exempt from the charges prescribed in this section.

SEC. 17. That all articles, goods, wares, or merchandise, imported into the Philippine Islands shall, for the purpose of this act, be deemed and held to be the property of the person to whom the same may be consigned; but the holder of any bill of lading, drawn to order and indorsed by the consignor, shall be deemed the consignee thereof; and in case of the abandonment of any article, goods, wares, or merchandise to the underwriters, the latter may be recognized as the consignee.

INVOICES.

SEC. 18. That all invoices of imported articles, goods, wares, or merchandise, shall state the true value thereof in the currency of the place or country from whence imported, or, if purchased, in the currency actually paid therefor, shall contain a correct description of such articles, goods, wares, or merchandise, with true numbers, weights and quantities, in the tariff terms of this act, and shall be made in quadruplicate, and signed by the owner or shipper, if the merchandise has been actually purchased, or by the manufacturer or owner thereof, if the same has been procured otherwise than by purchase, or by the duly authorized agent of such purchaser, manufacturer or owner.

SEC. 19. That except in case of personal effects accompanying a passenger as baggage, or arriving within a reasonable time before or after the owner, no importation of any articles, goods, wares, or merchandise, exceeding one hundred dollars in dutiable value, shall be admitted to entry without the production of a duly certified invoice of the kinds hereinafter described, or the filing of an affidavit made by the owner, importer, or consignee, before the collector of customs, showing why it is impracticable to produce such invoice, together with a bond in an amount to be prescribed by, and with sureties satisfactory to, the collector of customs, for the production of such invoice within a reasonable time to be prescribed by said official. In the absence of such invoice, no entry shall be made upon the aforesaid affidavit unless the same be accompanied by a statement in the form of an invoice, or otherwise, showing the actual cost of such merchandise if same was purchased, or if obtained otherwise than by purchase, the actual market value or wholesale price thereof at the time of exportation to the Philippine Islands in the principal markets of the country from whence imported. This statement shall be verified by the oath of the owner, importer, consignee, or agent desiring to make the entry, taken before the collector of customs, and it shall be lawful for that official to examine the deponent under oath regarding the source of his knowledge, information or belief, concerning any matter contained in his affidavit, and to require him to produce any correspondence, document, or statement of account in his possession, or under his control, which may assist the customs authorities in ascertaining the actual value of the importation or of any part thereof, and in default of such production when so required, such owner, importer, consignee, or agent shall be thereafter debarred from producing any such correspondence, document, or statement for the purpose of avoiding the imposition of additional duty, penalty, or forfeiture incurred under this or any other act in force in the Philippine Islands, unless he shall show to the satisfaction of the court or the collector of customs, as the case may be, that it was not in his power to produce the same when so demanded; but no articles, goods, wares, or merchandise shall be admitted to entry under the provisions of this section unless the collector of customs shall be satisfied that the failure to produce the required invoice is due to causes beyond the control of the owner, importer, consignee or agent.

Sec. 20. That invoices required by the preceding section shall, at or before the shipment of the merchandise, be produced to the consul, vice-consul, or commercial agent of the United States of the consular district in which the merchandise was manufactured or purchased, as the case may be, when importation into the Philippine Islands is from a country other than the United States of America, or any territory or place under the jurisdiction and control of the Government thereof: *Provided*, That the insular collector of customs may, in his discretion, dispense with the requirement for the consular invoices prescribed in this section in case the merchandise for which entry is sought is free of duty under this act, in which event a commercial invoice certified by the purchaser, manufacturer, seller, owner, or agent shall be filed; and *Provided further*, That when the importation is from the United States of America, or any territory or place under the jurisdiction and control of the Government thereof, production shall be to a collector of customs, deputy collector of customs, or United States commissioner.

Invoices shall have endorsed thereon when produced as above prescribed, a declaration signed by the purchaser, manufacturer, seller, owner, or agent, setting forth that the invoice is in all respects correct and true, and was made at the place from whence the merchandise is exported to the Philippine Islands; that it contains, if the merchandise was obtained by purchase, a true and full statement of the time when, the place where, the person from whom the same was purchased, and the actual cost thereof, and of all charges thereon; and that no discounts, bounties, or drawbacks are contained in the invoice except such as have been actually allowed thereon; and when obtained in any other manner than by purchase, the actual market value or wholesale price thereof, at the time of exportation to the Philippine Islands, in the principal markets of the country from whence exported; that such actual market value is the price at which the merchandise described in the invoice is freely offered for sale to all purchasers in said markets, and that it is the price which the manufacturer, seller, owner, or agent making the declaration would have received and was willing to receive, for such merchandise sold in the ordinary course of trade in the usual wholesale quantities, and that it included all charges thereon; that the numbers, weight, or quantity stated is correct, and that no invoice of the merchandise described, differing from the invoice so produced has been or will be furnished to anyone.   If the merchandise was actually purchased, the declaration shall also contain a statement that the amount shown and the currency stated in such invoice is that which was actually paid for the merchandise by the purchaser.   Said declaration shall be duly sworn to by the purchaser, manufacturer, owner, or agent, before the officer to whom produced.

SEC. 21. That consuls, vice-consuls, commercial agents, collectors, of customs, deputy collectors of customs, and commissioners of the United States of America, having any knowledge or information of any case or practice by which any person obtaining verification of any invoice, defrauds or may defraud the revenue of the Philippine Islands, shall report the facts to the insular collector of customs.

SEC. 22. That United States Government vessels, whether transports of the Army or naval vessels, when coming from the United States or a foreign port to the ports of the Philippine Islands, shall be subject to the same inspection by Customs Officers of the Philippine Government, for the purpose of determining whether they have on board articles of merchandise dutiable under the laws of the Philippine Islands, as such United States Government vessels are subject to by Customs Officers of the United States Government when such vessels enter ports of the United States from foreign countries for the purpose of determining whether such vessels have on board articles or merchandise dutiable under the laws of the United States.

DRAWBACKS.

SEC. 23. That on all fuel imported into the Philippine Islands which is afterward used for the propulsion of vessels engaged in trade with foreign countries or between ports of the United States and the Philippine Islands, or in the Philippine coastwise trade, a refund shall be allowed equal to the duty imposed by law upon such fuel, less one per centum thereof, which shall be paid under such rules and regulations as may be prescribed by the insular collector of customs.

SEC. 24. That upon imported materials, used in the manufacture or production of articles in the Philippine Islands, or in the packing, covering, putting up, marking, or labeling of articles grown, produced, or manufactured therein, and upon which the required amount of duty has been paid, there shall be allowed a drawback of the corresponding proportion of the amount so paid, less one per centum thereof, upon proof satisfactory to the collector of customs of the exportation of the whole or part of any importation upon which duty has been paid as aforesaid, used in any of the ways above set forth: *Provided*, That when the articles exported, or the coverings thereof, are in part of materials grown or produced in the Philippine Islands, the imported materials or the parts made from such materials shall so appear in the completed articles or packages that the quantity or measure thereof may be ascertained: *And provided further*, That the imported materials for which drawback is claimed under this section shall be identified, and the quantity of such materials used and the amount of duties paid thereon ascertained, if necessary; the facts of their use in manufacture, production, or packing in the Philippine Islands, and their exportation therefrom, determined, and the refund, if made, shall be paid to the manufacturer, producer, or exporter, to the agent of any of them, or to the person to whom such manufacturer, producer, exporter, or agent shall, in writing, order such refund paid, under and in accordance with such rules and regulations as the insular collector of customs may prescribe: *Provided, however*, That no drawback shall be paid under this section on account of any articles, goods, wares, or merchandise exported to the United States of America, or to any territory or place under the jurisdiction or control of the Government thereof, wherein such articles, goods, wares, or merchandise are admitted free of duty or at any rate of duty lower than the minimum rate of duty applicable to articles, goods, wares, or merchandise imported therein from foreign countries.

SEC. 25. That containers, such as casks, large metal, glass, or other receptacles, which are, in the opinion of the collector of customs, of such a character as to be readily identifiable, may be delivered to the importer thereof, upon identification and the giving of a bond with sureties satisfactory to the collector of customs in an amount equal to double the estimated duties thereon, conditioned for the exportation thereof, or payment of the corresponding duties thereon, within one year from the date of importation, under such rules and regulations as the insular collector of customs shall prescribe.

SEC. 26. That the index and repertory hereto attached are not an integral part of this act, and shall not be construed to alter or change the same in any manner.

SEC. 27. That the insular collector of customs shall, subject to the approval of the secretary of the department having jurisdiction over the customs service, make all rules and regulations necessary to enforce the provisions of this act.

SEC. 28. That all existing decrees, laws, regulations, orders, or parts thereof, inconsistent with the provisions of this act, except as provided in section one of this act, are to that extent repealed.

SEC. 29. That this act shall be known and referred to as the Philippine tariff revision law of nineteen hundred and nine.

SEC. 30. That this act shall take effect upon its passage.

PHILIPPINE ISLANDS.

	1905. Par.	1909. Par.
Abaca (*see* Hemp):		
Export duties on	398	355
Import duties	Class 5, Sec. II	Class VI
Waste for export		355
Abacus, Chinese, of wood	195, 196, 197	188
Abandonment of imported merchandise		Sec. 17
Abbreviations used in tariff		Sec. 7
Abrastol	99	80, 81
Absinthe, liqueur	308 (c)	263
Herb (wormwood)	81	57
Oil of	99	87 (a)
Absorbent cotton, not medicated	99 (a)	82
Abutment crane	247	194
Acacia gum	78	56 (b)
Accordions	195 to 197	185
Accumulators, electrical	250	193 (a)
Acetate:		
Lead	95 (a)	75
Methyl	97	77
Uranium	99	81
Zinc	95 (a)	75
Acetates (*see* Note)	95 (a)	75
Acetic acid	92 (c)	71 (b)
Acetic ether, for anesthetics	99	81
Acetic ether, commercial (not prepared for anesthetics)	97	77
Acetic anhydride, commercial	97	77
Aceton	97	77
Acetone	97	77
Acid:		
Acetic	92 (c)	71 (b)
Anhydride sulphuric	97	77
Benzoic	92 (d)	71 (b)
Boric	91 (a)	70 (a)
Camphoric	92 (d)	71 (b)
Carbolic	92 (a)	71 (a)
Carbonic (liquid)	91 (b)	70 (b)
Chromic	91 (c)	70 (c)
Citric	92 (a)	71 (b)
Fluoric	91 (c)	70 (c)
Gallic	92 (d)	71 (b)
Hydrobromic	91 (c)	70 (c)
Hydrochloric	91 (a)	70 (a)
Hydrocyanic	92 (d)	71 (b)
Hydrofluoric	91 (c)	70 (c)
Lactic	92 (d)	71 (b)
Muriatic	91 (a)	70 (a)
Nitric	91 (a)	70 (a)

	1905. Par.	1909. Par.
Agricultural tools and implements (not machinery):		
Axes	46	194
Cane knives	46	194
Grafters	46	194
Machetes	46	194
Mattocks	46	194
Pruning knives and shears	54 (e)	44 (a)
Rakes	46	194
Scythes	46	194
Sickles	46	194
Spades	46	194
Weeding hooks	46	194
Agua insalus	312 (c)	268
Agua salles (hair dye)	97	87 (b)
Air motors	243	194
Pumps	257	194
Airol	99	80–81
Alabaster:		
In the rough	1 (a)	1 (a)
In sculptures		1 (c)
		336, 337
In slabs or plates	1 (b)	1 (b)
In other articles	1 (d)	1 (c)
Alarm clocks	239	187
Albumen capsules, empty, for use in pharmacy	99	81
Albumen paper	187 (a)	151
Albumens	109	90
Albumenizing paper	187	151
Albums:		
Photo		151, 155
Of lithographs, etc		155
Of paper	190 (d)	151, (b), 155
Of leather	229	182
Alcanna root	81	57
Alcohol	306	260
Amylic	97	77, 260
Methyl	307	260
Wood	307	260
Alcoholmeters		303
Ale (malt)	312	267
Ginger	312 (c)	268
Aleuritis oil	100 (b)	83
Aleuritis tribola oil	100 (b)	83
Alimentary pastes for soups	284	225
Alimentary substances		Class XIII
Alkalies; caustic, barilla, potash	93	72
Alkaloids	96	78
Of cinchona bark	384	322
In pills, except those of quinine and cinchona alkaloidal salts	98	80, 81
Salts (except of cinchona alkaloids)	96	78
Almighty oil (Chinese)	99	80
Alloys of metals. (*See* corresponding metals.)		
Alloys of metals not mentioned (*see also* Zinc; etc)	73	54
Allspice:		
Unground	297 (a)	250 (a)
Ground	297 (b)	250 (b)
Oil of	99	87 (a)
Almond:		
Essential oil of bitter	105 (a)	87 (a)
So-called artificial oil of bitter (oil of mirbane)	97	87 (a)
Almond oil, expressed	99, 100 (b)	83
Almonds:		
Dried, in natural state, not shelled	329	239
Oil of sweet, commercial	100 (b)	83
Oil, expressed, refined	99	83

	1905. Par.	1909. Par.
Almonds—Continued.		
Essence of bitter, for perfumery	105 (a)	87 (a)
Extract of, for flavoring food	327	258, 263
Shelled	365	239
Aloe (fiber):		
Manufactures of		Class VI
Raw	138	313
Yarn of	142, 143	119
Textiles of. (*See* Textiles.)		
Thread, twine or cord	144	120
Rope or cordage	146	120
Ropemakers' wares	146	120
Aloes:		
Pharmaceutical extract of	99	81
Gum (bitter aloes)	78	56 (b)
Aloin	99	81
Alpaca (wool):		
Wool	162	137
Yarns	163	138
Textiles	166	142
Alpargatas, sandals	277 (c)	180 (b)
Alum	94 (c)	74 (c)
Ammonium ferric	94 (h)	74 (c)
Chromic	94 (h)	74 (c)
Aluminum:		
Chloride of	94 (h)	74 (c)
Cooking utensils	71 (d)	52 (b)
Lumps or ingots	71 (a)	309
Paint, aluminum powder in oil	84 (d)	61 (d)
Pipes	71 (b)	52 (a)
In powder	71 (e)	52 (b)
Sheets, bars, or wires	71 (b)	52 (a)
Other articles	71 (c)	52 (b)
Alumnol	99	81
Amarog sulfuroso	99	80
Amber:		
Wrought	341 (b)	280 (b)
Unwrought	341 (a)	280 (a)
Imitation of	342	281
Oil of	99	87 (a)
Amethyst:		
Cut	24	13
In rough	379	13
Amidol	97	77
Amidopyrine	99	81
Ammeters	248	193 (a)
Ammonia aqua	93	73
Anhydrous, liquefied	97	73
Borated bath, containing soap	104	86
Spirits of	99	81
Water	93	73
Ammonium:		
Bromide of	97	77
Carbonate	94 (e)	74 (c)
Chloride	94 (e)	74 (c)
Iodide	99	81
Sulphate of	94 (d)	74 (a)
Nitrate	94 (h)	74 (c)
Sulphocyanide	97	77
Valerianate	99	81
Ammunition		91
Amorphous phosphorus	90 (b)	69
Amperemeters	248	193 (a)
Amylic alcohol	97	77, 260
Amyl nitrate	99	81
Amygdalae amarae, essential oil of	105 (a)	87 (a)

	1905. Par.	1909. Par.
Analgesine	99	81
Anchor plugs and receptacles, electrical	248	193
Anchors	42	310
Anchovies:		
Canned	318	215 (c)
Salted or pickled	274 (b)	215 (c)
		216 (b)
Sauce of	328	215 (c)
Anemonina	99	81
Anethi, oil	105 (a)	87 (a)
Anethol	105 (a)	87 (a)
Angelica root	81	57
Angora goat, hair	161	314
Textiles of	164	139
Angulas (canned)	318	215 (c)
Anhydride sulphuric	97	77
Anhydride santonine, acid	99	81
Anhydrous ammonia	97	73
Aniline colors (dyes)	87 (b)	66 (c)
Aniline oil	97	77
Animal:		
Carbon (bone-char)	89	88
Fats	101	84
Oils	101	84
Product employed in medicine	82	59
Wastes—		
Unmanufactured	234	184 (a)
Manufactured		184 (b)
Wax—		
Crude	102	85 (a)
In candles	103	85 (b)
In other articles	103	85 (c)
Animals:		
Asses	207	170, 333
		342, 348
		334
Bovine, work		171 (a)
		333, 348
Breeding		
Bulls and cows	208 (b)	333, 342
		171 (a), 348
Calves, sucking	208 (c)	171 (b), 341
Mules	206	170, 334
Other, live	211	174
Oxen	208 (a)	171 (a), 334
Poultry, live	268	175
Sheep and goats	211	174
Singing birds and parrots	212	175
Swine	209, 210	172
Suckling pigs		173
Specimens of, prepared for museums, etc	389	344
Stallions and mares	205	170, 333
Stuffed or mounted	232	289, 290
Trained, for public entertainment	392	342
Trained, for exhibition		348
Anise, essence of	105 (a)	87 (a)
Anise seed	81	57
Oil, essential	105 (a)	87 (a)
Anisette	308 (c)	263
Anklets, knit:		
Of cotton	125 (c)	107 (c)
Of other vegetable fiber	152 (c)	127 (b)
Of wool	165 (b)	141 (c)
Of silk	173 (b)	148
Annunciators, electric	248	193 (a)
Annatto	85	66 (b)

	1905. Par.	1909. Par.
Anodyne		
Anthemidis oil	99	79, 81
Antiseptic surgical dressings	99	87 (a)
Antifebrine	99	82
Antifriction bushings. (According to material.)	99	80, 81
Antimony:		
Black sulphate of		
Crocus	94 (h)	74 (c)
Sulphide	97	77
Oxide, prepared for pharmaceutical use	94 (h)	74 (c)
Metallic powder	99	81
Metal. (See Lead; Zinc; etc.)	73 (f)	54 (b)
Antipyrine		
Anvils, wrought iron	99	81
Apiol	43	38
Apioline pills and capsules	99	81
Apocodeine	98	81
Aphomorphine	96	78
Apparatus	96	79
Agricultural		Class XII
Electrical	245	194
Gymnasium	248	193
For dredging		297
For hoisting	245	194
For making and repairing roads	245	194
For making sugar	245	194
For photography	245	194
For pile driving		302
For preparing rice, hemp, or other insular vegetable	245	194
products for the market	245	194
For refrigerating and ice making	245	194
For weighing	242	192
Philosophical and scientific, when free	390	349
Scientific, such as aerometers, barometers, compasses, quadrants, microscopes, theodolites, telescopes, thermometers, and the like	365	303
Apparel, finished, cut, basted, or partially finished, making up, surtax for		Rule 11, Sec. 5
Apparel, used, wearing, imported by passengers in their baggage	386	339, 340, 341, 347
Apple butter	322 (b)	237
Appliances for sight testing		303
Apricot oil	105 (a)	87 (a)
Aqua ammonia	93	73
Aqua regia	91 (a)	70 (a)
Arabic gum	78	56 (b)
Arachis oil	100 (b)	83
Arc lamps, and fittings for	248	193 (a)
Archæological objects for public museums, etc	388	344
Arexina, tannate of	99	81
Argentum nitricum	94 (h)	74 (c)
Argirol	99	80, 81
Argols		76
Argonin	99	80, 81
Argyrol	99	80, 81
Aristol	99	80, 81
Armor plate	42	31 (a)
Arms. (See Sec. 6.)		
Fire, of all kinds, and detached parts therefor	55, 56	46
Side arms, not fire, and detached parts therefor	54 (b)	44 (b)
Cartridges for; also caps and primers (see Sec. 6)	346	91
Aromatic seeds	81	57
Arresters, electric, lightning	248	193
Arrowroot	282	221

	1905. Par.	1909. Par.
Arrowroot flour	282	221
Arsenic:		
Fowler's solution of	99	80, 81
Iodide of	99	81
Pure, in lumps or powder	97	77
Arsycodile	99	81
Art, fine, works of:		
Destined for public museums, galleries, or art schools	387	336, 337, 344
Drawings, not for use in manufactures or industrial arts	380	326
Drawing for use in manufactures of industrial arts	182	155
Paintings		326
Articles:		
Falsely stamped or marked, of gold or silver or their alloys, prohibited		Sec. 6 (d)
Not otherwise provided for		306
Artificial:		
Buds	350	292
Christmas trees		296
Colors	84, 85	61
Dyes	87	66 (c)
Eyes and limbs		282
Fertilizers	106	74 (a)
Flowers	350	292
Fruits	350	292
Leaves		292
Limbs and eyes		282
Moss		292
Pistils		292
Parts of fruits and flowers of whatever material		292
Seeds		292
Stone	2	2
Teeth, with plates or not	342 (b)	282
Artists' colors	85	61
Asafœtida gum	78	56 (b)
Oil of	99	87 (a)
Asarum oil	99	87 (a)
Asbestos, in any form	2, 3	4
Aseptic surgical dressings	99	82
Ash, soda		72
Asperin	99	81
Asphalt, also paving blocks of	7	22
Asphalt paints	85 (a)	61 (b)
Aspirators	54 (e)	196, 303
Aspirin	99	80, 81
Asses	207	170
Asthma cigarettes	99	81
Astrakhan (according to material)	166	
Astronomical appliances and apparatus not otherwise provided for		303
Atropine	96	78
Atropine sulphate	96	78
Attachment plugs, electric	248	193
Augers	46	194
Ships	46	194
Bits	46	194
Aurantii corticis, essential oil of	105 (a)	87 (a)
Aurantii florum, oil of	105 (a)	87 (a)
Automatic slot machines	256	Sec. 6, 186, 191
Automatic brakes, vacuum	257 (b)	194
Automobiles and detached parts therefor	261, 352	198
Avena	282	221
Awls	46	194
Axes	46	194
Axle grease of all kinds	8, 9	22

	1905. Par.	**1909.** Par.
Axles of wrought iron	38	33
Wood, for wagons (*see* note)	266	202
Babbits lye	93	72
Babbit metal. (*See* Tin alloy.)		
Baby carriages (perambulators)	261	201
Baby rattles:		
Silver	28 (d)	26 (d)
As toys	353	296
Bacon:		
Canned	315, 316	209
In bulk	270	207
Badges, when free		330
Bagatelle balls		166
Bagatelle, pool, and billiard tables, parts and appurtenances for		166
Baggage (travelers'), when free	386	338, 340, 341, 347
Bags:		
Gunny	145	121
Of paper	178	150, 151
Straw, fine	204 (a)	169 (c)
Tool, of whatever material		288
Traveling, of whatever material	228 (c)	288
Baked beans in cans	320	233
Baking powder	97	77
Balm oil	105 (a)	87 (a)
Balances of all kinds for weighing	242	192
Baling presses	245	194
Balloons and parts for		201
Of rubber, for children	353	296
Balls:		
Bagatelle		166
Billiard	341, 342	166
Bowling		166
Pool	341, 342	166
Of all kinds, for use in sports (except bagatelle, billiard, bowling, and pool balls)	229	297
For children, toys	353	296
Balsam of:		
Copaiba	99	81
Fir, Canada	78	56 (b)
Peru	99	81
Tolu	78	56 (b)
Balsams:		
In a natural state	78	56 (b)
Prepared in capsules	98	81
Otherwise prepared	99	81
Bamboo brooms	204 (a)	169 (c)
Bamboo:		
In straight lengths for parasols and umbrellas	343	169 (a)
In furniture	204	169 (b)
Leaves	203	169 (a)
Manufactured into combs	204 (a)	169 (c)
Manufactured into fans	338	277
Manufactured into other articles	204 (a)	169 (c)
Shoots, edible	288	231, 232
Unwrought	203	169 (a)
Banana flowers:		
Dried or in cans	288	231, 232
Oil, artificial	97	87 (a)
Bandages, surgical, sterilized	99	82
Band saws	245, 275 (b)	194
Bands:		
Hats—		
Leather	229	182
Of paper	190 (d)	151 (b)

	1905. Par.	1909. Par.
Bands—Continued.		
Of oilcloth	349 (d)	287 (b)
Rubber (soft)	352 (d)	293 (b)
Cotton, for cinches and saddle girths	134 (d)	116
Jute, for cinches and saddle girths	159 (a)	134
Lithographed paper for tobacco, etc	183	151 (b)
Barbed wire, wrought iron	45	36 (a)
Barbers' chairs, of whatever material	199, 59	165
Bar fixtures, wood	199	162, 163, 164
Barilla alkalies	93	72
Barium:		
Chlorate	94 (h)	74 (c)
Chloride	94 (h)	70 (c)
Nitrate	94 (h)	74 (c)
Sulphate, natural, as color	83	60
Bark:		
For dyeing and tanning	86 (a)	66 (a)
Used in medicine	81	57
Soap bark	81	57
Barley:		
In grain	277 (a)	219 (a)
In flour	277 (b)	219 (a)
Pearl	282	221
Barometers	365	303
Barrels, casks, tuns, and pipes, water-tight	363	161 (a)
Staves, hoops and headings for	194 (b)	160
Not water tight	194 (a)	161 (b)
For firearms	55 (b)	45
Barrows, wheel; of whatever material, and detached parts therefor	266	197
Bars of:		
Aluminum	71 (b)	52 (a)
Cast-iron	31 (a)	28 (a)
Copper	62	48
Gold	372	318
Lead	73 (b)	54 (a)
Nickel	71 (b)	52 (a)
Platinum	372	318
Silver	372	318
Steel, ordinary	35 (b)	30 (b)
Crucible	35 (c)	30 (c)
Cut to measure	41	34
Tin	72 (b)	53 (a)
Zinc	73 (b)	54 (a)
Other metals	73 (b)	54 (a)
Barytes		60
Baseball:		
Bats	195	297
Gloves	220	179 (b)
Baseballs	229	297
Basic photographic paper	187	151
Basils, curried	217 (a)	178 (a)
Basins. (According to material.)		
Bas-reliefs:		
Of gypsum	4 (a)	7
Of marble, jasper, alabaster, or similar fine stone	1 (c)	1 (c)
Of clay, faience, stoneware, porcelain, or bisque	23	11 (f)
Bathing suits. (According to material.)		
Bathing trunks, knit:		
Cotton	125 (c)	107 (a)
Linen	152 (c)	127 (b)
Wool	165 (b)	141 (c)
Silk	173 (b)	148
Bath robes, pile fabrics:		
Of cotton	124 (a)	106
Of linen	151	126

	1905.	1909.
	Par.	Par.
Bath towels. pile fabrics:		
Cotton	124 (a)	106
Linen	151	126
Bath tubs; according to material		Sec. 12
Baths, developing		302
Bats, baseball, or cricket	195	297
Batteries:		
Galvanic	248	193 (b)
Dry	248	193 (a)
Medical induction	54 (e)	193 (b)
Storage	249	193 (a)
Wet	248	193 (a)
Battery zincs	248	193 (a)
Batting, of cotton	112	98
Baydean laffecteur	99	80, 81
Bay:		
Oil of	99	87 (a)
Rum	195 (b)	87 (b)
Bead curtains, glass	16	20 (e)
Beads:		
Glass	16	20 (e)
In necklaces	340	279
Imitating amber	342 (b)	281, 282
Wooden	195, 196, 197	162, 163, 164
In necklaces	340	281
Beam scales	242	192
Beams:		
Cast iron	31 (a)	28 (a)
Steel—		
Ordinary	35 (b)	30 (b)
Crucible	35 (c)	30 (c)
Structural		34
For weighing	242	192
Wooden, ordinary	192 (a)	157
Bean curd	321	233
Beans:		
Canned	320	232
Castor	81	57
Cacao	290 (a)	248 (a)
Dried—		
In bulk	287 (a)	229 (a)
In small packages	287 (b)	229 (b)
Flour of	287 (c)	229 (c)
For drugs	81	57
Fresh	375	230
St. Ignatius	81	57
St. John's (Johannis Brod)	81	57
Tonka	81	57
Vanilla	295	259
Bearings, brass or copper	67	194 (b)
Beberine	96	78
Bed linen, used, imported by passengers in their luggage	386	339, 340. 341, 347
Bedspreads. (According to material.)		
Bedsteads. (According to material.)		
Beef:		
Canned or potted	315 or 316	209
Extract of	365	213
Fresh	376	205
In brine or salt	269	206
Iron and wine (medicine)	99	80
Jerked	269	207
Beer:		
Malt	312 (a), (b)	267
Root	312 (c)	268

	1905. Par.	1909. Par.
Beets:		
Fresh	375	230
Canned or potted	320	232
Pickled	321	233
Bees, live	365	174
Beeswax	102, 103	85
Bell metal (free)		309
Belladona:		
Herb	81	57
Pharmaceutical preparations of	99	81
Bellows:		
Blacksmith	257 (b)	194
Other. (According to material.)		
Belting, machine (of whatever material)		195
Other. (According to material.)		
Belts:		
Electric	365	193 (b)
Others. (According to material.)		
Bench lathes	257 (b)	194
Bench stops	46	194
Benedictine	308 (c)	263
Ben oil	99, 100 (b)	83
Bensol, chemically pure	97	77
Bent wood, furniture of	198	162
Benzaldehyde	97	77
Benzine	10	23 (b)
Benzoate of soda	99	81
Benzoic acid	92 (d)	71 (b)
Benzoin gum	78	56 (b)
Benzol	97	77
Benzolnaphtol	99	81
Berberine	96	78
Bergamot oil	105 (a)	87 (a)
Bergoin oil	105 (a)	87 (a)
Berlin blue, in powder	84 (c)	61 (c), (d)
Berries:		
For alimentary purposes—		
Fresh	374	234
Dried or desiccated	286	235
Canned or potted	322	236
Pickled	321	238
For dyeing	86 (a)	66 (a)
For drugs	81	57
Juniper	81	57
Betol	99	81
Betula oil, volatile	105 (a)	87 (a)
Beverages:		
Fruit juices or sirups for		258, 269
Malt	312	267, 268
Nonalcoholic	312 (c)	268, 269
Spiritous, compounded	308	263, 264
Bibles and extracts therefrom	382 (b)	328
Bibulous paper	187	150, 151
Bicarbonate of potassium	94 (h)	74 (c)
Bicarbonate of sodium	94 (g)	74 (c)
Bichloride of mercury	94 (h)	74 (c)
In tabloids	98	81
Bichromate of potash	94 (h)	74 (c)
Bicycles and detached parts for	252	199
Cement	109	90
Bigota, liquid extract of	99	80, 81
Bi-iodide of mercury	99	81
Billheads of paper	179	151
Billiard, pool, and bagatelle tables	200	166
Balls		166
Chalk	5	8
Cloth	166	142
Parts and appurtenances for	200	166

	1905. Par.	1909. Par.
Bills:		
Of exchange, executed		320
Of lading		Sec. 17
Binnacles		310
Biographies	254	186
Birch oil	105 (a)	87 (a)
Birds:		
Eggs—		
Fresh, salted, or preserved in natural form	377, 333	272
Preparations of		272 (b)
Live, including poultry	212	175
Nests, edible	366	226
Specimens for museums	389	344
Stuffed or mounted	232	289, 290
Biscuit:		
Unsweetened	283 (a)	223 (a)
Sweetened	283 (b)	223 (b)
Shredded wheat	282	221
Bismae	99	81
Bismuth:		
Metallic (metal not mentioned)	73	54
Salts of	99	81
Bisque	23	11
Bisulphate of quinine	384	322
Bisulphate of sodium	94 (h)	74 (c)
Bitartrate of potassium	97	76
Bit braces	46	194
Bits, wrought iron, for horses	49	42
Bit stork drills	46	194
Bitter almond oil:		
Essential	105 (a)	87 (a)
So-called artificial (oil of mirbane)	97	87 (a)
Bitters of all kinds for use with beverages	308 (c)	263
Bitumens	7	22
Bituminous paints	85 (a)	61 (b)
Blackberry brandy	308 (b)	262
Blackboards of hyloplate	190 (d)	151
Blacking, shoe	89	67
Blacksmith's bellows	257 (b)	194
Bladders		183
Blades, concealed, weapons with		44 (c)
"Blanco," for shoes	89	67
Blank books	181	153
Blankets, according to material.		
Blasting gelatine	111 (a)	91 (a)
Blasting powder	111 (a)	91 (a)
Blind, books and music in raised print for		328
Blocks:		
Pulley, differential	245	194
Snatch, steel	58	47
Others. (According to material.)		
Blondes. (According to material.)	186	150, 151
Blotting paper		
Boards of:		
Cork	202 (a)	168 (a)
Common wood	192	157
Fine wood	193	158
Boas, feather	230	290
Boats	267	203
		Class XII
		203
Importation of, defined (see note)		
Boiled rice	282	221
Boiled cider	313	267
Boiler coverings of asbestos	2 (c)	4
Boilers:		
Cooking utensils. (According to material.)		
Steam	244	194

	1905. Par.	1909. Par.
Bologna sausages, not canned	270	207
Bolts, expanding (tools)	46	194
Bolts of:		
Wrought iron or steel	47	39
Copper or brass	68	50 (a)
Bond paper	179	150, 151
Bone:		
Black	89	61
Buttons	345 (a)	283 (b)
Char		88
Cuttle fish	366	184
Meal for fertilizing		184 (a)
Shaved, for Chinese medicine	99	80
Wrought	342 (b)	281 (b)
Unwrought	342 (a)	281 (a)
Bones, animal wastes	234	184
Bonnets of:		
Straw, chip, palm, grass, rattan, osiers	355	299
Other materials	356	300
Books:		
Blank	181	153
Bound or unbound, printed	180	154
Letterpress, copying	187	153
Of lithographic prints	180	155
Novels	180	154
Printed and written matter. objects obscene, indecent, or subversive of public order, prohibited	Sec. 6	Sec. 6
Philosophical, historical, scientific	390 (a)	349
Raised print, for the blind	390 (b)	328
Reimported		350
School, text	382 (b)	328
Scientific, imported for established institutions	390 (a)	349
Used, imported by passengers in baggage	386	339, 340, 341, 347
Boot laces, according to material.		
Boot and shoe findings		178
Boots and shoes of whatever material	221-225	180
Bootstraps:		
Of cotton		113
Of linen		131
Borax	94 (e)	74 (c)
Boric acid	91 (a)	70 (a)
Borers, tap	46	194
Boring machines	257 (b)	194
Boron	90 (b)	69
Bort		13
Botany, specimens of, for public museums	389	344
Bottles:		
Glass--		
Common	12	15 (a)
Fine, for toilet purposes	13	16
Other	16	20
Clay--		
Plain	19 (c)	11 (b)
Painted	19 (d)	11 (e)
For household ornaments or in vases	23	11 (f)
Bovine animals	208	171 (a), 333, 334, 342, 348
Bowling:		
Alleys, parts and appurtenances for		166
Balls		166
Bowls for opium pipes	23	Sec. 6
Box cars	262 (b)	200 (a)
Boxes (see Rule 13):		
Of cardboard—		
Common, plain	190 (c)	150 (b)
Fancy	190 (d)	151 (b)

	1905. Par.	1909. Par.
Boxes—Continued.		
Of cardboard—Continued.		
Not fancy, covered with surface-coated paper...	190 (b)	151 (b)
Cedar wood boards for cigar....................	192 (b)	157 (b)
Hat, of whatever material......................	228 (c)	288
Lubricating, for railway trucks, and carriages—		
Of cast iron....................................	31 (b)	28 (b)
Of wrought iron or steel......................	38 (a)	33 (a)
Of wood—		
In which imported merchandise is regularly packed......................................	192 (c)	Rule XIII, 355
Retail..	195, 196, 197	162, 163, 164
Wooden boards (common) planed for..............	192 (b)	157 (b)
Boxing gloves..	220	179 (b)
Braces and bits..	46	194
Brads...		41
Braid, according to material.		
Braids, admixtures of materials in....................		Rule 7, Sec. 5
Brake beams (railroad)...............................		33 (a)
Brake shoes (railroad)...............................		33 (a)
Brakes, vacuum, automatic...........................	257 (b)	194
Bran...	303	228
Brandied fruits.......................................	323	238
Brandies...	308 (a)	261
Blackberry and ginger...........................	308 (b)	262
Fruits preserved in (brandied fruit).............	323	238
Brass. (See Copper.)		
Brass bushings, bearings............................	67	50, 194
Brass screws..	68	50
Brass, old...	60	309
Brazing compound, chemical.........................	97	77
Bread:		
Sweetened....................................	283 (b)	223 (b)
Unsweetened..................................	283 (a)	223 (a)
Bread knives..	54 (a)	44 (a)
Breeding animals, when free..........................		333
Bricks:		
Asphalt, for paving.............................	7	22
Common clay, of all kinds.......................	17	9
Bridges, iron and steel bars for.....................	41	34
Bridles of leather....................................	228	181 (b)
Brine, meat in.......................................	269	206
Brimstone (sulphur)..................................	90 (a)	68
Bristles...	161	139, 314
Brushes, textiles, and other manufactures of.......	164	139
Bristol board in sheets...............................	189 (a)	150, 151
Britannia metal (see Tin, alloys of).................	72	53
Broche, or brocaded textiles........................	Rule 9	Rule 8, Sec. 5
Broches:		
Definition of...................................		Rule 8, Sec. 5
Cotton, with silk, surtax on....................		Rule 8, Sec. 5
Of vegetable fibers other than cotton; with silk.....		Rule 8, Sec. 5
Bromhydrate of hyoscyamine.........................	96	78
Bromide of:		
Ammonium....................................	97	77
Potassium.....................................	97	77
Sodium.......................................	97	77
Bromine..	90 (b)	69
Bromoform...	99	81
Bromo seltzer..	99	81
Bromo soda...	99	81
Bronze powder..	69 (a)	50 (a)
Bronze vases...	69	50
Bronzed hides and skins..............................	219	178 (c)
Bronzing liquid (methyl acetate).....................	97	77
Broom corn...		169

	1905. Par.	1909. Par.
Brooms:		
Bamboo, palm, fine straw, and similar materials....	204 (a)	169 (c)
Basswood, street.	204	169 (c)
In which bristles (or hair) are chief value	164	139
Brushes:		
Bristle.	164	139
Camel's hair.	164	139
Carbon.	248	193
Rabbit's feet mounted, as for toilet purposes.	365	290
Rotary steel wire.	257 (b)	194
Steel foundry.	46	194
Fine straw, bamboo, palm, or similar materials.	204 (a)	169 (c)
Other. (According to material.)		
Brussels carpet.	166	142
Buchu leaves.	81	57
Buckets, according to material.		
Buckles, as trinkets or ornaments.		279
Other. (According to material.)		
Bucksaws.	46	194
Buckthorn sirup.	99	81
Buckwheat:		
In flour.	278 (b)	218 (b)
In grain.	278 (a)	218 (a)
Budding knives.	54 (c)	44 (a)
Buds, artificial.	350	292
Buds for drugs	81	57
Bulbous roots for drugs.	81	57
Bulbs for drugs.	81	57
Bullets, lead, for firearms.	73 (b)	54 (b)
Bulls, live cattle.	208 (b)	171 (a), 333, 334, 342, 348
Bulrushes.	203, 204	169
Bumpers (railroad).		33 (a)
Bungs, hoops, headings, shooks, and staves		160
Buoys, life.		310
Burglar alarms, electric.	248	193
Burgundy, pitch.	77 (a)	56 (a)
Burgundy, sparkling wine.	309	264
Burnishers, tools of steel.	46	194
Burnt umber (ocher).	83	60
Bushings, antifriction. (According to material.)		
Bushings, brass bearings.	67	50, 194
Bushings, electric.	248	193
Butcher's:		
Cleavers.	46	194
Knives.	54 (a)	44 (a)
Tools.	46	194
Butter.	335	274
Fruit.	322 (b)	237
Cacao.	290 (b)	248 (b)
Ghee.	336	275
Imitation of.	336	275
Butterine.	336	275
Button cards:		
Cut out, punched, or perforated.	190 (a)	150 (b)
The same further manufactured or elaborated.	190 (d)	151 (b)
Button fasteners.		278
Button rings.		278
Buttons.	345	283
Buzzers, electric.	248	193 (a)
By-products, animal.		184
Cabinetmakers, fine wood for.	193	158
Cabinet organs.	236	185
Cables:		
Copper, for conducting electricity.	65	49 (c)
Wrought iron or steel.	45	36 (a)
Submarine telegraphic.	395	323
Hawsers of hemp.	146 (b)	120 (a)

	1905. Par.	1909. Par.
Cacao:		
Bean	290 (a)	248 (a)
Butter	290 (b)	248 (b)
Ground, in paste or powder	290 (b)	248 (b)
Cade collodion	99	81
Cade oil	99	87 (a)
Cadium oil	99	87 (a)
Caffeine	96	78
Cajuput oil	99	87 (a)
Cakes, canned	325	224
Calamus oil	105 (a)	87 (a)
Calcaria glycerophosphorica	99	81
Calcimines, substances prepared for		60
Calcium:		
Carbide	94 (e)	74 (c)
Carbonate, precipitate (chalk)	3 (a)	6 (b), (c)
Glycerophosphate	99	81
Hypochlorite		74 (b)
Hypophosphite	99	81
Iodide	99	81
Lactophosphate	99	81
Lactate	99	81
Phosphate, precipitated	94 (d)	74 (c)
Sulphide	94 (h)	74 (c)
Calendars:		
Advertising, when free	381	327
Printed	190 (d)	151
Lithographic	190 (d)	151
Calfskin:		
Curried	217 (b)	178 (b)
Shoes of	223	180 (b)
Calipers, steel	46	194
Calisaya:		
Elixir of	99	80, 81
Bark	81	57
Call bells, electric	248	193 (a)
Calomel	99	81
Caltrop nuts	329	239
Calves, suckling	208 (c)	171 (b)
Calves' foot jelly	332	211
Camel's hair	161	139, 314
Brushes, textiles and other manufactures of	164	139
Cameos	24	13, 280
Cameras and parts thereof	358–59–60	302
Camphor	78	56 (b)
Ice	99	81
Oil of	99	87 (a)
Camphoric acid	92 (d)	71 (b)
Canada fleabane oil	99	87 (a)
Candies	332	244, 245
Candles, wax	103	85 (b)
Cotton wicks for	116 (b)	96
Candlesticks:		
Glass	13	16
Other. (According to material.)		
Cane packing	204 (a)	Rule 13
Cane:		
Knives	46	194
Furniture	204	169 (b)
Cut in straight lengths	343	169 (a)
Unwrought	203	169 (a)
Other manufactures	204 (a)	169 (c)
Split or stripped		169 (d)
Canned:		
Angulas	318	215 (c)
Bacon	315, 316	209
Breads	324	223

	1905. Par.	1909. Par.
Canned—Continued.		
Cakes	325	224
Codfish	317	215 (a)
Clam broth	319	212
Cream	320 (a)	270
Fish—		
Common	317	215 (b)
Delicatessen	318	215 (c)
Fruit	322	236
Branded or pickled	321, 322, 323	238
Herring	317	215 (a)
Meats	315, 316	209, 210, 211
Milk	320	270
Meluza Guisada	318	215 (c)
Oysters	317 (b)	215 (b)
Frozen	275	217
Puddings	325	224
Salmon	317 (a)	215 (a)
Sardines	317	215 (a), (c)
Soup	319	212
Sweetmeats	325	244
Vegetables	320	232
Canoes	267	203
Can openers	46	194
Cantharides:		
Whole or powdered	82	59
Tincture of	99	81
Canvas shoes	221	180 (a)
Lacquered, in imitation of patent leather	222	180 (b)
Caoutchouc (rubber):		
Belting, for machinery	352 (e)	195
Boots and shoes	352 (c)	180 (b)
Gaskets	352 (a)	293 (a)
Hard rubber articles	352 (b)	293 (c)
Hose	352 (e)	294
Other articles of soft rubber	352 (d)	293 (b)
Packing for machinery	352 (a)	293 (a)
Raw	77 (c)	293 (a)
Sheets	352 (d)	293 (a)
On textiles of—		
Cotton	135 (a)	117
Linen	160	135
Wool	166	142
Silk	175	148
Washers	352 (a)	293 (a)
Capers:		
Fresh	375	230
Pickled	321	233
Sauce	328	256
Caps:		
For wearing apparel	357	301
For miners, explosive		91 (a)
Percussion		91 (b)
Capsules:		
Albumen, for medicine, empty	99	81
Apioline	98	81
Gelatin, for medicine, empty	99	81
For bottles of—		
Aluminum	71 (e)	52 (b)
Lead	73 (f)	54
Tin	72 (e)	53 (b)
Nickel	71 (c)	52 (b)
Zinc		54
Medicinal	98	81
Carabana water	312 (c)	268
Carabao		171 (a)
When free		334

	1905. Par.	1909. Par.
Caraway oil	99	87 (a)
Caraway seeds	81	57
Ground for culinary purposes	296 (b)	255 (b)
Carbide of calcium	94 (e)	74 (c)
Carbolic acid	92 (a)	71 (a)
Carbolineum	7	22
Carbonate of:		
Ammonia	94 (e)	74 (c)
Iron, saccharated	99	81
Magnesia	94 (c)	74 (c)
Nickel	94 (h)	74 (c)
Potassium	94 (h)	74 (c)
Sodium	94 (e)	74 (c)
Zinc	94 (h)	74 (c)
Carbonated waters	312 (c)	268
Carbonating machinery	257	194
Carbon brushes for electricity	248	193
Carbon dioxide	91 (b)	70 (b)
Carbonic acid, liquid	91 (b)	70 (b)
Carbon paper	187 (c)	151
Carbons	110	193 (a)
Carboys, glass	12	15 (a)
Cardboard:		
Manufactures of	190	150 (b), 151 (b)
Sheets	189	150, 151
Traveling bags	228 (c)	288
Matches	351	92
Card mounts for photos	190	150, 151
Cards:		
Index	190 (a)	150, 151
Marked		Sec. 6 (c)
Playing	353	296
Visiting	190 (a)	150, 151
For buttons		150, 151
Carlsbad, salt of, natural	99	81
Carlsbad sprudel water	312 (c)	268
Carmine	85	61 (c), 61 (d)
Carob:		
Balsam	99	81
Beans	302	57
Essential oil of	99	87 (a)
Carpets. (According to material.)		
Carriage:		
Apron studs or knobs of steel	58 or 59	47
Harness, of leather	228 (b)	181 (b)
Lamps. (According to material.)		
Carriage tires of rubber	352 (d)	293 (b)
Carriages, of kinds, including perambulators	258–263	201
Cars:		
Freight	262, 263	200 (a)
Other	262, 263	200 (b)
Carton-pierre (papier mache)	191	156
Cartridges, loaded or not, for firearms	346	91 (b)
Carts and wagons	265, 266	197
		296
Toy, and other small vehicles for children's use	197	164
Carved wood		
Carylopsis:		
Oil of	105 (a)	87 (a)
Toilet perfumery of	105 (b)	87 (b)
	99	81
Cascara sagrada	98	80
Cascarets		
Cases:		303
For apparatus and instruments	239	187
For clocks	192 (c)	Rule 13, 355
Common wood, for the packing of imported goods		288
Suit, of whatever material		

	1905. Par.	1909. Par.
Cases—Continued.		
Telescope	204 (c)	288
Tool, of whatever material		288
Cashmere:		
Hair of	161	139, 314
Textiles	166	139
Cash registers	255	189
Casks, barrels, tuns, and pipes	363	161
Cassia oil	105 (a)	87 (a)
Cassimeres	166	142
Castings of steel or malleable iron	37, 38	32, 33
Cast iron, malleable, articles of, are dutiable as wrought iron.		
Castile soap	104	86
Castoreum	99	81
Castor oil	100 (b)	83
Catch-all		306
Catgut:		
Ligatures, aseptic	99	82
Strings for musical instruments	233	185
Catheters, surgical instruments	54 (e)	196
Catnip leaves	81	57
Cats, live	211	174
Catsup	328 (a)	256
Cattle, live	208	348, 171, 333, 334
Caustic alkalies	93	72
Caustic pencils	99	81
Cauteries, thermo	54 (e)	193 (b)
Caviar, canned	318	215 (c)
Cayenne pepper	299	252
Cedar woods:		
In boards for cigar boxes	192 (b)	157 (b)
Logs	192 (a)	157 (a)
White, oil of	105 (a)	87 (a)
Yellow, oil of	99	87 (a)
Celery:		
Canned	320	232
Fresh	375	230
Pickled	328	233
Salt	328	256
Sauce	328	256
Oil	99	87 (a)
Seed ground for culinary purposes	296 (b)	255
Celloidina	97	77
Celluloid, or imitations of:		
Unwrought	342 (a)	281 (a)
Wrought	342 (b)	281 (b)
Films for cameras	362	302
Films for cinematographs and similar machines	254	186
Cement:		
Bricks, squares, tiles, and pipes	3 (a) 17, 18	9
Bicycle (glue)	109	96
Dentists'	97	77
Industrial	3 (a)	6 (a)
Portland	3 (a)	6 (a)
Stain	89	61
Centrifugal machines	257	194
Cerae oil	99	84
Ceramic products		Class I
Ceramic tiles:		
Plain	18 (a)	10 (a)
Glazed or decorated	18 (b)	10 (b)
Ceramyl (a sizing compound)	97	77
Cereals:		
Prepared for table use	282	221
Not elsewhere provided for	278	219

	1905. Par.	1909. Par.
Cerium:		
Valerianic	99	81
Salicylate of	99	81
Ceylon cinnamon oil	105 (a)	87 (a)
Chaff cutters	245	194
Chain hoists	245	194
Chains; as trinkets or ornaments		279
Others. (According to material.)		
Chairs:		
Barbers' or dentists'	32, 59, 199	165
Of cast iron, railway	31 (b)	28 (b)
Of wrought iron, railway	38 (a)	33 (a)
Of bent wood	198	162
Of common wood	195	162
Of fine wood	195	163
Of fine wood	196	164
Of wood, gilt, carved, etc	197	164
Chafing dishes. (According to material.)		
Chalcedony (semiprecious)		13
Chalk:		
Crude	3 (a)	6 (b)
Billiard, French, red, or tailors'	5	8
Precipitated		6 (c)
Chamois leather	219	178 (c)
Chamomile oil	99	87 (a)
Champagne	309	264
Chandeliers, electric		193 (b)
Other. (According to material.)		
Char, bone		88
Charcoal:		
Animal (bone char)	89	88
In fuel	201	167
In powder or tablets (pharmaceutically prepared)	99	81
Willow, for medicine	99	81
Chartreuse	308 (c)	263
Charts	182	155
Chatterton compound, electric insulating material	248	193 (a)
Checks and drafts:		
Executed		Free, 320
Other (blank forms)	179, 181, 183	151
Cheese	334	273
Substitutes for		273
Chemical:		
Dye colors	87 (b)	66 (c)
Fire engine. (According to material.)		
Fertilizers	106	77 (a)
Products (not otherwise provided for)	97	77
Chenopodium oil	97	87 (a)
Cheroots, tobacco	364 (b)	304 (e)
Cherry juice	313	269
Wild	327	258
Chemical industries; substances employed in		Class IV
Cherries:		
Fresh	374	234
Preserved	322 (b)	236
Maraschino, brandied	322, 323	238
Chessmen. (According to material.)		
Chest protectors. (According to material.)		288
Chests, tool, of whatever material	332	244
Chewing gum	315, 316	210, 211
Chicken, canned or potted	268	175, 204
Chickens	292	246, 247
Chicory, in any form		
Chimneys for lamps:		
Of glass	13 (c)	17
Of porcelain	21	11
Of mica	2 (e)	5

	1905. Par.	1909. Par.
China:		
Clay (kaolin)	3 (b)	6 (b)
Wares	23	11
Chinchona:		
Oil	99	87 (a)
Bark	81	322
Alkaloids	384	322
Chinese:		
Almighty oil	99	80
Bird's nest, edible	366	226
Joss money	190 (d)	150, 151
Joss sticks	105 (b)	87 (b)
Lanterns (paper)	190 (d)	151
Lily nuts	366	239
Lucky paper	187	150, 151
Medicinal pills	98	80
Medical preparations and materials	99	80
Marking ink	85	64 (b)
Mulberry paper	187	150, 151
Pillows, covered with leather	229 (b)	182 (b)
Plasters	99	81
Rock sugar	298 (b)	240 (c)
Shoes and boots	225, 227	180
Wines—		
Containing 14 per cent or less of alcohol	310, 311	265
More than 14 per cent, not more than 24 per cent of alcohol	310, 311	266
More than 24 per cent	308 (c)	263
Chin, solution of	99	81
Chinolin or quinoline	384	322
Chirurigal plaster	99	81
Chisels	46	194
Chloralhydrate	99	81
Chlorate of:		
Barium	94 (h)	74 (c)
Iron	94 (h)	74 (c)
Potassium	94 (f)	74 (c)
Sodium	94 (f)	74 (c)
Chloride of:		
Ammonium	94 (e)	74 (c)
Calcium	94 (h)	74 (c)
Gold	96	78
Lime (bleaching powder)	94 (e)	74 (b)
Magnesium	94 (h)	74 (a)
Mercury	99	81
Platinum	96	78
Potassium	94 (c)	74 (a)
Silver	96	78
Sodium	94	74 (c)
Uranium	99	81
Zinc	94 (h)	74 (c)
Chloridine	99	80, 81
Chlorinated lime	94 (e)	74 (c)
Chloroform	99	81
Chlorhydrate of herion	96	78
Chlorhydrate of quinine	384	322
Chlorplatinate of potassium	97	77
Chocolate:		
For manufacturing purposes	330 (a)	249 (a)
For table use	330 (b)	249 (b)
In candies, sweetmeats, etc	332	244
Choke coils for use with car motors	250	193 (a)
Christmas-tree ornaments	353	296
Christmas trees, artificial		296
Chrome alum	94 (h)	74 (c)
Chromic acid	91 (c)	70 (c)

	1905. Par.	1909. Par.
Chromolithographs	183	151, 155
Chronometers and parts of	241	187
Chrysarobin	99	81
Chucks	46	194
Church regalia, gems and statuary	393 (b)	336
Churns	257	194
Chutney sauce	328	256
Cicutine	96	78
Cider	312, 313	267
Cigarette holders. (According to material)		
Machines	257 (b)	194
Paper	188	152
Cigarettes	364 (b)	304 (e)
Cigarettes, cubeb	99	81
Cigar:		
Box labels	183	151
Boxes, cedar wood, boards planed for	192 (b)	157 (b)
Lighters, electric	248	193 (e)
Cigars	364 (b)	304 (b)
Cinches:		
Of cotton	134	115
Of hemp	159	133
Cinchona:		
Bark	81	322
Alkaloids or salts of	384	322
Ferruginous sirup of	99	81
Oil of	99	87 (a)
Cinchonidine, and salts thereof	384	322
Cinchonine sulphate	384	322
Cinematographs, records and parts for	254	186
Cinnamon	293, 294	250
Oil of	105 (a)	87 (a)
Circus equipments, imported temporarily, bond taken for re-exportation	392	342
Ciridinia	99	81
Citrate:		
Of ammonium	99	81
Of magnesia	99	81
Of sodium, not pharmaceutically prepared	95 (b)	75
Of iron, not pharmaceutically prepared	95 (b)	75
Citrates	95 (b)	75
Citric acid	92 (a)	71 (b)
Citron, conserved or crystallized	331	238
Citronilla oil	105 (a)	87 (a)
Citrophen	99	81
Civette:		
Essential preparation for manufacturing perfume	105 (a)	87 (b)
In toilet perfumery and handkerchief extracts	105 (b)	87 (b)
Clam:		
Broth and chowder	319	212
Shells	366	284
Clams:		
Canned or in glass	317 (b)	215 (b)
Other	275	216 (b)
Claret wines	311	265, 266
Clasp nails. (See Nails.)		
Classification:		
Advance, opinion concerning, how to obtain		Rule 12, Sec. 5
Advance, opinion concerning, not to be given in ad- vance except as provided		Rule 12, Sec. 5
Clay:		
Common	3 (b)	6 (b)
China (kaolin)	3 (b)	6 (b)
Bricks	17	9
Bottles	19 (c)	11 (b)

	1905. Par.	1909. Par.
Clay—Continued.		
Dishes	19 (b), (d)	11
Flower pots, common	19 (c)	11 (b)
Flower stands, statuettes, vases	23	11 (f)
Household and kitchen utensils	19 (a)	11
Tubes or pipes	17 (c), (d)	9
Pigeons	19	11 (c)
Tiles	18	9
Cleats, porcelain, for electrical wiring	248	193 (a)
Cleavers, butchers'	46	194
Clippers, hair	54 (d)	44 (b)
Clocks, and parts therefor	239, 246, 241	187
Cloisonne wares	23	11 (f)
Cloth, emery	46	194 (b)
Clorhidro fosfato	99	81
Clothing:		
Finished, cut, basted, or partially finished, making up, surtax for		Rule 11, Sec. 5
Ready-made, ascertainment of threads		Rule 1, Sec. 5
Clover (forage)	303	228
Seed	302	227
Cloves	296	250
Oil of	105 (a)	87 (a)
Coaches	258	201
Coal	6	21
Drawback on		Sec. 25
Coal tar	7	22
Colors, derived from	87 (b)	66 (c)
Creosote, crude	7	22
Cobalt colors	84	61 (c), (d)
Cocaine	96	78
Cochineal	86 (e)	66 (c)
Cocktails	308 (b)	263
Cocoanut:		
Oil	100 (a)	83
Shredded	332, 365	239
Cocoanuts	76	55
Export duties on	402	358
Fiber		Class VI
Cocodylate of iron	99	81
Cocoons, silk	168	332
Codein	96	78
Codfish:		
Canned or potted	317	215 (a)
Salted or dried	273	214
Intestines, part of	273	183
Cod-liver oil:		
Emulsion, and other pharmaceutical preparations of.	99	81
Other	101	84
Coffee	291 (a), (b)	245
Coffins. (According to material.)		
Cognac:		
Oil of	105 (a)	87 (a)
Brandy	308 (a)	263
Coins of national issues (free)		320
Coir	138	Class VI
Hawsers, rope, and fenders	146	120
Mats and carpeting	155	135
Waterproofs	204 (a)	169 (c)
Coke	6	21
Colchiflor	99	81
Colcothar	97	77
Cold cream	105 (b)	87 (b)
Collargol blister	99	81
Collars. (According to material.)		
Collections of botany, mineralogy, and zoology, when free.	389	344

	1905. Par.	1909. Par.
"Colliers royer"	341 (b)	286 (b)
Collodion	97	77
Cantharidal	99	81
Cologne, perfumery prepared for toilet purposes	105 (b)	87 (b)
Colophony	77 (a)	56 (a)
Colts		170
Coloring for wine and liquors	85	61 (c), (d)
Colors:	87 (b)	66 (c)
Aniline	87 (b)	66 (c)
Artists'	85	61 (c), (d)
Artificial	84, 85	61 (c), (d)
Berlin blue	84 (c)	61 (c), (d)
Bituminous	85 (a)	61 (b)
Boneblack	89	61 (c), (d)
Cochineal	86 (c)	66 (c)
Derived from coal tar (aniline)	87 (b)	66 (c)
Coal tar paint, not aniline dyes	85 (a)	61 (b)
Carmine	85	61 (c), (d)
Lithographic (ink)	85 (a)	64 (a)
Litmus	58 (a)	61 (c), (d)
Logwood dyeing extracts	87 (a)	66 (b)
Litharge	84 (c)	61 (c), (d)
Lampblack and ivory black	89	61 (c), (d)
Lead—		
Red or white, dry	84 (a)	61 (a)
In liquid or paste	84 (b)	61 (b)
Natural colors—		
Dry	83	60
In liquid or paste	84 (d)	61 (d)
Noncorrodible graphite	89	67
Ocher—		
Dry	83	60
In liquid or paste	84 (d)	61 (d)
Prussian blue	84 (c)	61 (c), (d)
Sugar (burnt)	85	61 (c), (d)
Sienna—		
Dry	83	60
In liquid or paste	84 (d)	61 (d)
Ultramarine blue	84 (c)	61 (c),)d)
Umber—		
Dry	83	60
In liquid or paste	84 (d)	61 (d)
Venetian red	83, 84 (c), (d)	61 (c), (d)
Vermilion	84 (c)	61 (c), (d)
Water	85	61 (c), (d)
White zinc	84 (c), (d)	61 (c), (d)
Columns:		
Of cast iron	31 (a)	28 (a)
Of wrought iron or steel (structural)	41	34
Combed silk floss	171 (b)	146
Combs, curry	58, 59	47
Combs. (According to material.)		320
Commercial documents, executed (free)		335
Commercial samples	370	
Common woods:		
Acacia, alder, ash, beech, birch, black poplar, California redwood, cedar, cypress, elder, evergreen oak, maple oak, pear, pine, plantain, poplar, spruce, yew-leaved fig		157
	46	194
Compasses for measuring	365	303
Compasses, magnetic		Rule 12, sec. 5
Component material of chief value defined	345 (a)	283 (b)
Composition buttons	272	208
Compound lard		

	1905. Par.	1909. Par.
Compounds:		
Explosive	111 (a), (b)	91
Flavoring		258
Insulating, used exclusively for electricity	248	193 (a)
Milk		271
Compressed air:		
Engines	243	194
Pumps	257	194
Comptographs, and detached parts for		188
Computing apparatus, and parts for		188
Concentrated milk		270
Condensed milks and creams	320	270
Condiments for table use	328	256
Cones of glass for paving or roofing		18 (a)
Confectionery	331, 332	244
Confetti paper	190 (d)	150
Coniine	96	78
Conserved or crystallized fruits	331	238
Consignees		Sec. 17
Construction and enforcement		Sec. 5
Consular invoices	Sec. 22	Sec. 20, 21
Consular supplies, official, when free		345
Containers and receptacles:		
Immediate, of imported merchandise		Rule 13 (d), sec.5
Reexportation of		Sec. 25
Controllers, electrical	250	193 (a)
Cooked rice	282	221
Cooking soda	94 (g)	74 (c)
Coopers' wares suitable for use as containers of liquids	194	161 (a)
Other	194	161 (b)
In shooks, hoops, headings, staves, and bungs	194 (b)	160
Copaiba:		
Balsam	99	81
Oil of	105 (a)	87 (a)
Copal varnish	88	62
Copper, and alloys of, in:		
Bars	62	48
Bearings	67	50
Bolts	68 (a), (b)	50
Buttons	345 (b)	283 (c)
Bronze powder	69 (a)	50 (a)
Coins of national issues	Sec. 19 (c), 69 (a)	320
Cables, electric	65	49 (c)
Foil	69 (a)	50
Ingots (alloys)	61	48
Ingots (pure)	61	309
Laminæ (scales)	60	309
Matte		309
Muntz metal		309
Nails—		
Nickeled	68 (a)	50 (b)
Other	68 (b)	50 (a)
Old	60	309
Ores	369	309
Pens	68 (c)	278
Pins	68 (c)	278
Pipes	67	48
Plates for fireplaces	67	50
Regulus		309
Rivets	68	50
Sheets	63	48
Scales (laminæ)	60	309
Screws	68 (a), (b)	50
Wind instruments (musical)	69 (a), (b)	185
Washers	68 (a), (b)	50
Tacks	68 (a), (b)	50
Wire	64, 65, 66	49

	1905. Par.	1909. Par.
Copper, and alloys of, in—Continued.		
Other manufactures of—		
Plain	69 (a)	50 (a)
Nickeled	69 (b)	50 (b)
Strings for musical instruments	69 (a), (b)	185
Oxide of	94 (e), 99	74 (c)
Nitrate of	94 (e)	74 (c)
Sulphate of	94 (e)	74 (c)
In pencils for cauterization	99	81
Copra	76	55
Export duties on	402	357
Copying books, letterpress	187	153
Copying paper	187	150
Coral:		
Unwrought	341 (a)	280 (a)
Wrought	341 (b)	280 (b)
Compositions imitating	342	281
Cord. (According to material.)		
Cordage. (According to material.)		
Cordials	308 (c)	263
Coriander seeds	81	57
Ground for culinary purposes	296 (b)	255
Coriander oil	99	87 (a)
Corduroy (cotton)	124	103
Cork:		
Rough or in boards	202 (a)	168 (a)
Manufactured in stoppers	202 (b)	168 (b)
Manufactured in other articles		168 (c)
Corkscrews	46	194
Corn:		
Broom		169
Canned	320	232
In grain	278 (a)	220 (a)
In meal or flour	278 (b)	220 (b)
Shellers	245	194
Cornstarch, prepared for table use	282	221
Correspondence (free)		349
Corrugated rubber matting	352 (d)	293 (b)
Corrugated sheets of wrought iron or steel, galvanized or not	36 (c)	31 (b)
Corrosive sublimate	97	77
Corset laces. (According to material.)		
Corset stays. (According to material.)		
Corsets. (According to material.)		
Corticine		287
Corundum or carborundum wheels	46	3
Cosmetics	105 (b)	87 (b)
Cosmos starch	107	89
Cot beds. (According to material.)		
Cottolene	272	208
Cotton, and manufactures thereof		Class V
Absorbent, medicated or not	99	82
Bath robes, pile fabrics	124 (a)	106
Bathing suits, knitted	125 (c)	107 (c)
Batting		98
Bedspreads of lace	127 (a)	109
Blankets	123	104
Blondes	127	109
Boot straps		113
Braid	131	113
Carpeting	128	110
Cinches	134	115
Cord, wrapping	116 (a)	96
Cordage	133	96
Corduroy	124	103
Corset laces	132	114

Cotton, and manufactures thereof—Continued.	1905. Par.	1909. Par.
Curtains—		
Of tapestry	129	111
Of lace	127	109
Other. (According to material.)		
Drawers, knitted	125 (b)	107 (b)
Elastic textiles	135 (b)	117
Embroidery	Rule 10	Rule 9
Felt		98
Galloons	131	113
Gloves, knit	125 (c)	107 (c)
Gins	245	194
Hammocks		97
Jerseys, knit	125 (b)	107 (c)
Knitted goods—		
In the piece	125 (a)	107 (a)
Corset covers	125 (c)	107 (c)
Fishing nets	144	96
Gloves	125 (c)	107 (c)
Hammocks	144	97
Shawls	125 (c)	107 (c)
Stockings and socks	125 (c)	107 (b)
Sweaters	125 (b)	107 (b)
Tennis nets	144	97
Undershirts and drawers	125 (b)	107 (b)
Vests, ladies'	125 (b)	107 (b)
Laces	127	109
Lace bedspreads, curtains, and pillow shams	127 (a)	109 (a)
Mackintoshes	135 (a)	117
Medicated	99	82
Mops		98
Paper stock		315
Pile fabric	124	105
In towels	124 (a)	106
In bath robes	124 (a)	106
Pillow shams of lace	127 (a)	109
Piques	122	103
Plush	124	105
Raw	112	312
Ribbon	131	113
Rope	133	96
Rugs	128	110
Saddle girths	134	115
Seed oil	100 (b), 305	83
Seed	76	55 (a)
Seed meal or cakes		55 (b)
Shoe laces	132	114
Socks, knitted	125 (c)	107 (b)
Stockings, knitted	125 (c)	107 (b)
Swabs		98
Table covers of tapestry	129	111 (b)
Tape	131 (a)	113 (a)
Tapestry	129	111
Tassels	131	113
Tennis nets		97
Thread—		
For sewing, crocheting, embroidering, or darning	116	95
For sewing sails or sacks	116 (a)	96
Textiles	Class V, Group 3	
Velvets and velveteens	124	105
Vests, ladies' knit	125 (b)	107 (b)
Waste	112	93
Waterproof	135 (a)	117
Wicks—		
For lamps	130	112
For making candles or matches	116 (b)	96
Yarn	113, 114, 115, 116	94
Mercerized	113 (b), 116	95

	1905. Par.	1909. Par.
Cough drops	99	81
Counterpanes. (According to material.)		
Making up defined, surtax for		Rule 11, Sec. 5
Couplings		33 (a)
Coverings, unusual, for imported merchandise, pay duty at the same rate as would be charged if imported separately. (Sec. 177, Act No. 355)		Rule 13 (h)
Coverings (packing), when free		Rule 13
		355
Cow hair	161	139, 315
Cowhide shoes	221	180 (a)
Cowhides:		
Raw, green, or dried	214	176
Tanned with hair on	215	177
Tanned and curried	217 (d)	178 (a)
Varnished, grained, or embossed	218	178 (c)
Cows	208 (b)	171 (a), 333, 334, 342, 348
Crackers:		
Edible	283	223
Fire	111	91 (c)
Cranes, power, hand, or hydraulic	247	194
Cranks, wrought iron	38 (b)	32, 33, 194
Cravenettes	166	142
Crayon pictures of actual persons	380	326
Other	182	155, 326
Crayons:		
Chalk	5	8
Charcoal, for drawing	85 (c)	65
In wood as pencils	85 (c)	65
Cream of tartar	97	76
Cream, toilet	105 (b)	87 (b)
Cream ware. (*See* Stoneware.)		
Creams and milks	320	270
Creme de menthe	308 (c)	263
Cremor bismuth	99	81
Creolin	97	77
Creoline (unrefined creosote)	7	22
Creosote:		
Refined (wood)	99	81
Unrefined	7	22
Crepe paper	187	150, 151
Crezol	99	81
Cricket bats		297
Crochet hooks	53, 68 (c)	278
Crochet lace:		
Cotton	127	109
Linen	154	128
Silk	174	148
Croquet sets and parts of		297
Croton oil	99	83
Crowbars	46	194
Crucibles of clay or stoneware	19 (b)	11 (b)
Crucible steel:		
Bars	36 (c)	30 (c)
Tools	46	194, 196
Crude:		
Animal oils and fats	101	84
Oils, mineral	8	22
Materials not otherwise provided for	366	306
Petroleum	8	22
Rules for classification of		Class II, Group 2
Crushers, rock	245	194
Crystal, and glass imitating crystal	13–16	16, 20
Artificial eyes	16 (c)	282
Chandeliers or bracket lamps	13 (a), (b)	16

	1905. Par.	1909. Par.
Crystal, and glass imitating crystal—Continued.		
Cut, engraved, painted, enameled, or gilt...........	13 (a)	16 (b) 20 (c)
Chimneys, for lamps.............................	13 (c)	17
Globes, for lanterns............................	13 (b)	16
Electric incandescent lamps......................	250 (a)	193 (a)
Field glasses..................................	16 (c)	20 (e)
In statuettes, flower stands, vases, and similar articles, for toilet purposes or house decoration..............	13 (a), (b)	20 (b), (c)
Mirrors.......................................	15	19
Plates, for roofing or paving.....................	14	18
Spectacle glasses..............................	16 (a)	20 (a)
Tinned, silvered, or coated with other metals (mirrors)....................................	15	19
Window......................................	14 (b), (c)	18 (b), (c)
Crystallized fruits................................	331	238
Crystals, for watches.............................	238 (a)	187
Cubebs..	81	57
Cigarettes of.................................	99	81
Oil of..	99	87 (a)
Cucurbitine......................................	96	78
Cuffs, celluloid..................................	342 (b)	281 (b)
Cultivators......................................	245	194
Cumarin...	99	81
Cups, as trophies or prizes, when free..............		330
Curacao, liquor..................................	308 (c)	263
Currants:		
Canned.......................................	322 (b)	236
Dried..	286	235
Currency, receivable for duties....................	Sec. 8	Sec. 9
Of invoice....................................		Sec. 18
Of national issue..............................		320
Currycombs, wrought iron.........................	58, 59	47
Curry powder....................................	296 (b)	255 (b)
Curtains. (According to material.)		
Cut glass, articles of.............................	13 (a), 16 (c)	16 (b) 20 (c), (e)
Cuticura ointment................................	99	81
Cutlery of steel..................................	54	44
Cutter circuit breakers............................	248	193 (a)
Cutters, forage..................................	245	194
Cutters, glass...................................	46	196
Cuttings of common metals, fit only for re-smelting......		309
Cuttlefish bone..................................	365, 366	184
Cyanide of potassium.............................	97	77
Cyanide of mercury...............................	97	77
Cyclometers, and parts therefor....................		187
Dana:		
Indian coriander seed..........................	81	57
Ground for culinary purposes...................	296 (b)	255 (b)
Darning cotton...................................	116	95
Dates:		
Dried—		
In bulk...................................	286	235 (a)
In small packages.........................	285	235 (b)
Fresh..	374	234
Canned.......................................	322 (b)	236
Davits..	245	310
Deals:		
Ordinary wood................................	192 (a)	157
Fine wood....................................	193 (a)	158
Decanters, glass.................................	13 (a), (b)	16 (a), (b)
Decorations, house:		
Of bisque, clay, faience, porcelain, and stoneware...	23	11 (f)
For Christmas trees...........................		296
Definitions of terms used in the tariff..............		Sec. 8
Deleterious articles prohibited.....................		Sec. 6 (e)

	1905. Par.	1909. Par.
Delphina, pure	96	78
Demijohns:		
Of clay	19 (b), (d)	11 (b)
Of common glass	12	15 (a)
Denaturants, mixtures of		76
Dental:		
Instruments	54 (e)	196
Rubber in sheets	352 (d)	293
Dentists:		
Chairs	59, 199	165
Cement	97	77
Gold solder	27 (d)	25 (d)
Silver for	28 (d)	26 (d)
Moss fiber gold	27 (d)	25 (d)
Dentifrices	105 (b)	87 (b)
Desk:		
Fans, electric	250	193 (a)
Telephones	248	193 (a)
Developing:		
Baths		302
Lights		302
Trays		302
Deviled ham, canned	315, 316	210
Dextrine	108	89
Dials for watches	238	187
Diamond-pointed tools of steel	46	196
Diamond dust	365	13
Diamonds	24, 379	13
Diastase	99	81
Dice. (According to material.)		
Loaded (prohibited)	Sec. 6	Sec. 6 (c)
Dies and stocks	46	194
Differential pulleys	245	194
Diggers, post hole	46	194
Digitalin	96	78
Digitalis (herb)	81	57
Dill, oil of	105 (a)	87 (a)
Diminutive articles for use as toys		296
Dingies	267	203
Dionine	96	78
Dioxide of manganese	97	77
Dishes. (According to material.)		
Disk harrows	245	194
Distributing boards, electric	248	193 (a)
Diuretina	99	81
Diving suits and appurtenances	257	194
Documents:		
Commercial, executed		320
Public		320
Typewritten		320
Dog biscuits	283 (a)	223
Dogs	211	174
Doilies. (According to material.)		
Domestic animals, when free		341
Dominos of:		
Wood or bone	342 (b), 353	296
Ivory	341 (b)	280
Donkey engines	243	194
Doorknobs. (According to material.)		
Doublets:		
Unset	24	14
In gold jewelry	27 (c)	25 (c)
In silver jewelry	28 (c)	26 (c)
Dover's powder	99	81
Downs and feathers		291

	1905. Par.	1909. Par.
Draft harness	228 (a)	181 (a)
Drafts (blank forms)		151 (b)
Executed		320
Dragees, medicinal	98	81
Dragon's blood	87 (a)	66 (c)
Drain pipes. (According to material.)		
Drawbacks on imported materials	Sec. 23	Sec. 23, 24
Drawbars (railroad)		33 (a)
Drawers. (According to material.)		
Drawing inks	85 (b)	64 (b)
Drawing paper	187	150, 151
Drawings in ink, water colors, or oil	182	155, 326
Dredging machinery and apparatus	245	194
Dressing:		
Harness	89	67
Hoof	99	81
Dressing for shoes	89	67
Dressings, surgical (aseptic and antiseptic)	99	82
Dress patterns of paper	190 (d)	151
Dress shields. (According to material.)		
Drill grinder, twisted	257 (b)	194
Drills	46	194
Drivers, screw	46	194
Drop black	89	61 (c)
Drugs		Class IV
Adulterated prohibited		Sec. 6 (e)
Fruits, flowers, and vegetable products for	81	57
Dried insects and animal products for	82	59
Drum hoists, friction	245	194
Dryers, tobacco	257	194
Dues, wharfage	Sec. 14	Sec. 16
Duplicating machines and parts for		188
Dusters, feather	231	291 (b)
Dust, gold, silver, and platinum	372	318
Dutiable:		
Value defined		Rule 13, (a), (b), Sec. 5
Weight		Rule 13 (c), (d), (e), (f), (g), (h)
Duties:		
Payment of		Sec. 9
Rates of, established		Sec. 11
To be assessed when two or more rates are applicable		Rule 12, Sec. 5
Duty, maximum rates of, on importations		Sec. 11
Dye, hair	97, 105 (b)	87 (b)
Dyeing extracts	87 (a)	66 (b), (c)
Dyes	87	66
Dye woods, barks, roots	86	66 (a)
Extracts of		66 (b), (c)
Dynamite	111 (a)	91 (a)
When prohibited	Sec. 6	Sec. 6
Dynamos	250	193 (a)
Earthenware. (*See* Clay.)		
Earth in kaolin	3 (b)	6 (b), (c)
Earths	3	Class I, Par. 6
Earth color, such as sienna, ocher	83	60
Eau de cologne	105 (b)	87 (b)
Eau de quinine	105 (b)	87 (b)
Eave troughs, galvanized sheet iron	36 (c)	31 (b)
Eaxine pink dye	87 (b)	66 (c)
Edam cheese	334 (b)	273
Edible bird's nest	366	226
Edible products not otherwise provided for:		
Crude		276 (a)
Other		276 (b)
Egg powders		272

	1905. Par.	1909. Par.
Eggs:		
Fresh, salted, or preserved in natural form............	333, 337	272 (a)
Preparations of..		272 (b)
Or silkworm..	167	332
Eggshell ware. (*See* Porcelain.)		
Eider down..	231 (a)	291 (a)
Manufactures of..	230	291 (b)
Elastic textiles. (According to material.)		
Electric:		
Amperemeters..	248	193 (a)
Annunciators...	248	193 (a)
Arc lamps and fittings.................................	248	193 (a)
Batteries—		
Dry and wet.......................................	248	193 (a)
Storage...	249	193 (a)
Bells...	248	193 (a)
Belts...	365	193 (b)
Buttons...	248	193
Carbon brushes...	248	193 (a)
Chandeliers...		193 (b)
Cigar lighters..		193 (b)
Conducting cables of copper.........................	55	49 (c)
Cooking and heating apparatus and utensils.........		193 (b)
Curling irons...		193 (b)
Dental appliances......................................		193 (b)
Desk and table lamps.................................		193 (b)
Dynamos..	250	193 (a)
Exciters...	250	193 (a)
Fans..	250	193 (a)
Fire-alarm apparatus...................................	248	193 (a)
Ferraris meters..	248	193 (a)
Flatirons..		193 (b)
Galvanometers..	248	193 (a)
Generators...	250	193 (a)
Incandescent bulbs and tubes.......................	250 (a)	193 (a)
Installations...	248	193 (a)
Insulating compounds.................................	248	193 (a)
Insulators of glass and porcelain....................	248	193 (a)
Lamps, incandescent..................................	250 (a)	193 (a)
Lighting, carbons for..................................	110	193 (a)
Motors..	250	193 (a)
Plating outfits...		193 (b)
Power, machinery for..................................	250	193 (a)
Snap switches...	248	193 (a)
Sockets...	248	193
Soldering irons..		193 (b)
Surgical appliances....................................		193 (b)
Switchboards..	248	193 (a)
Switches..	248	193 (a)
Stoves..	248	193 (b)
Tapes, insulating...	248	193 (a)
Telephones...	248	193 (a)
Therapeutic appliances................................		193 (b)
Thermostats...	248	193 (b)
Transformers..	250	193 (a)
Vibratory apparatus....................................		193 (b)
Voltmeters...	248	193 (a)
Wattmeters..	248	193 (a)
X-ray machines...		193 (b)
Other instruments, implements, utensils, and articles used for, by, or with.............................		193 (b)
Electricity:		
Apparatus and appliances.........................	248	193
Machinery for generation of.......................	250	193 (a)
Electro-plating outfits..............................		193 (b)
Electro silicon, metal polish..............................	97	77
Embossed hides, skins, and leathers....................	218	178 (c)

	1905. Par.	1909. Par.
Feathers:		
For ornaments	230	290
Plumes of	230	290
Others, and manufactures of	231 (a)	291
Feculae for industrial purposes	107	89
Felt:		
Caps of	357	301
Cotton		98
Covered with tar or pitch for roofs and structural purposes	348	286
Hats of	356	300
Wool	166	142
Fenalgina	99	81
Fencing:		
Foils		297
Iron or steel wire	45	36
Wooden	194 (d)	157 (b)
Other. (According to material.)		
Fennen oil	105 (a)	87 (a)
Fern oil	99	87 (a)
Ferraris meters	248	193 (a)
Ferricodile	99	81
Ferro cyanide:		
Of iron	97	77
Of potassium	97	77
Ferroline, aluminum powder in oil	84 (d)	61 (d)
Ferrum arsenate	99	81
Ferrum redact, pure iron for medicinal use	99	81
Fertilizers:		
Artificial or chemical	106	74 (a)
Natural (manures)	383	321
Fezzes		301
Fiber or pulp indurated	365	156
Fibers, dried for drugs	81	57
Fibers, vegetable, raw or hackled		313
Fichus, making up defined, surtax for		Rule 11, Sec. 5
Field glasses	16 (c)	20 (e)
Figs:		
Dried—		
In bulk	286	235 (a)
In small packages	285	235 (b)
Fresh	374	234
Figured hides, skins and leathers	218	178 (c)
Figures:		
Offensive to morality	Sec. 6 (2)	Sec. 6
Wax—		
For public entertainment (imported temporarily)	392	342
Other	103	85 (c)
Files:		
Letter, of cardboard	190 (d)	150 (b), 151 (b)
Tools	46	194
Filings of steel and other common metals fit only for re-smelting	74	309
Fillers, wood		62
Film packs, and kits		302
Films:		
For cameras	362	302
For cinematographs, biographs, and similar machines	254	186
Filter paper	187	150, 151
Findings of leather for boots and shoes	218	178 (a), (b), (c)
Fine art, works of, imported for public museums, galleries, or art schools	387	336, 337, 344

	1905. Par.	1909. Par.
Fine woods: Amaranth, apple, bird's-eye maple, box-wood, camphor, cherry, chestnut, ebony, hazel, holly, ironwood, jasmine, juniper, laurel, lemon, lignum-vitae, linden, mahogany, medlar, plum, pomegranate, orange, olive, rosewood, sandalwood, snakewood, teak, walnut, and yew..................................		158
Fine stones.........	1	1
Finishings, harness, of iron......................	49	42
Firearms, prohibited except as provided (*see* Sec. 6 (a)).	55, 56	45
Cartridges for and detached parts for.................	346	91 (b)
Fire:		
Alarm apparatus, electric...........................	248	193 (a)
Bricks of clay........................... 17 (a), (b)		9
Clay, articles of............... 17–19 (a), (b), (c), (d), (e), (f)		9, 11
Crackers................................	111 (b)	91 (c)
Engines...........	257	194
Extinguishers, chemical......................	97	77
Wood........................	201	167
Works..................................	111 (b)	91 (b)
Fire pans. (According to material.)		
Firmer chisels.....................................	46	194
Fish:		
Anchovies, canned................................	318	215 (c)
Angulas, canned...............................	318	215 (c)
Canned or potted...................................	317, 318	215
Caviar, canned................................	318	215 (c)
Clams—		
Canned................................	317 (b)	215 (b)
Not canned.............................	275	217
Codfish—		
Canned................................		215 (a)
Salted or dried....................................	273	214
Fresh......................	274 (a)	216 (a)
Herring, canned........................	317	215 (a)
Hooks, steel............................	54 (d)	44 (a)
Live..............................		174
Merluza guisada, canned.......................	318	215 (c)
Nets—		
Of cotton........................		96
Of other vegetable fibers.....................	144	120 (a)
Not otherwise provided for.........................	274	215 (b), 216
Oil..........................	101	84
Oysters—		
In bulk, or in cans, fresh	275	217
Preserved in cans.............................	317 (b)	215 (b)
Pickled, in bulk............................	274 (b)	216 (b)
Roe—		
Canned....................................	318	215 (c)
In bulk, dried...............................	274 (b)	216 (b)
Salmom, in cans...........................	317 (a)	215 (a)
Sardines, in cans.........................	317	215 (a), (c)
Salted, not canned........................	274 (b)	216 (b)
Shell, not canned or potted.......................	275	217
Smoked, not canned........................	274 (b)	216 (b)
Sounds..........................		183
Stockfish.........................	273	214
Fishing:		
Boats.......................	267	203
Nets, of cotton......................		96
Of hemp and other vegetable fibers..............	144	120 (a)
Fishplates...............................		33 (a)
Fittings for arc lamps.........................	248	193 (a)
Fittings of gold or silver plated wares for carriages and coffins......................		27 (b)
Flags. (According to material.)		
Flanges, malleable cast-iron..........................	58	47
Flaps for tobacco, etc., lithographed...................	183	151 (b)

	1905. Par.	1909. Par.
Flash light:		
Powder or sheets	73 (f)	54 (b)
Cartridges of magnesium	73 (f)	54 (b)
Flasks. (According to material.)		
Flatirons, of cast iron	32, 33	29
Flavoring extracts, compounds, and sirups	327	258
Flax:		
Cords and cordage	144, 146	120
Hackled	137	313
Manufactures of		Class VI
Plushes	151	126
Raw	137	313
Rope	146 (a)	120
Rope makers' wares	146	120
Seed	76	55 (a)
Ground, in meal	81	55 (b)
Oil of	100 (b)	83
Textiles, and manufactures of		Class VI, Group 2
Thread	144	120
Twine	144, 146	120
Velvets and velveteens	151	126
Yarns	139, 141	119, 120
Fleabane oil, Canada	99	87 (a)
Fleges (hoop iron)	36 (a), (c)	31 (b)
Flint glass. (See Crystal.)		
Flock (wool waste)	162	137 (a)
Flooring:		
Common wood planed or dovetailed for	192 (b)	157 (b)
Fine wood planed or dovetailed for	193 (b)	158 (b)
Florida water	105 (b)	87 (b)
Floss silks	171, 172	146
Flour:		
Arrowroot	282	221
Barley	277 (b)	219 (b)
Beans	287 (c)	229 (c)
Buckwheat	278 (b)	219 (b)
Cereals prepared for table use	282	221
Corn (maize)	278 (b)	220 (b)
Graham	277 (a)	219 (b)
Lily root	282	221
Millet	279 (b)	220 (b)
Oats	278 (b)	220 (b)
Other cereals	278 (b)	220 (b)
Pease	287 (c)	229 (c)
Pulse	287 (c)	229 (c)
Rice	276 (c)	218 (c)
Rye	277 (b)	219 (b)
Sacks. (According to material.)		
Wheat	277 (b)	219 (b)
Flowerpots:		
Clay and earthenware, common	19 (c)	11 (b)
Other, ornamental, of bisque, faience, porcelain, and stoneware	23	11 (f)
Flower seeds	302	227
Flower stands, decorative. (According to material.)		
Flowers:		
Artificial and parts of	350	292
Drugs	81	57
Lily, dried	288	231
Linden, dried	81	57
Rose	81	57
Fluoric acid	91 (c)	70 (c)
Fly paper, sticky or impregnated	97	77
Fodder	303	228
Foeniculi, oil of	105 (a)	87 (a)

	1905. Par.	1909. Par.
Foil:		
Copper	69 (a)	50
Gold	27 (d)	25 (d)
Gold or silver plated		27 (b)
Silver	28 (d)	26 (d)
Tin	72 (c)	53 (a)
Foils, fencing	59	297
Folders, for advertising only, having no commercial value	381	327
Footballs	229, 352 (d)	297
Forage..:	303	228
Cutters	245	194
Forceps, surgical and dental instruments	54 (e)	196
Other	46	194
Forking spades	46	194
Forks. (According to material.)		
Formaldehyde	97	77
Formalin	97	76
Formodine	99	81
Formula, private		Sec. 8
Formulary, national or pharmacoepia		Sec. 8
Fortoin	99	81
Fountain pens		295
Fowls (see Poultry)	268	175, 204
Frames for buildings of iron or steel	41	34
Frames, umbrella, steel	59	47, 298 (d)
Frames, plate		302
Frankfort drop back	89	61 (c), (d)
Frauds or intended frauds in invoices to be reported by consuls, vice-consuls, commercial agents, collectors, and deputy collectors of customs, and commissioners of the United States to the insular collector of customs		Sec. 21
Free (subject to express conditions)		Sec. 13
Free (upon compliance with corresponding regulations).		Sec. 14
Free list (unconditional)	Sec. 12	Sec. 12
"Freeze-em," a meat preservative	97	77
Freezers, ice cream	257 (b)	194
French chalk	5	8
Friction drum hoists	245	194
Frog, steel railway switch	42	33 (a)
Fruit:		
Artificial and parts of	350	292
Brandied	323	238
Butter		237
Canned	322	236
Conserved or crystallized	331	238
Currants, dried	286	235
Dried	285, 286	235
Fresh	374	234
Jams	322 (b)	237
Jellies	332	237
Juice, unfermented	313	268
Other		269
Nonedible, dried (drugs)	81	57
Oil, artificial	308 (c), 97	77, 263
Pickled	321	238
Brandied, in cordials or spirits	323	238
Preserved in wood, tin, or glass	321, 322	236
Pulp	313	237
Sirups for beverages	313	269
Fuel, vegetable	201	167
Furnaces:		
Clay	19 (b)	11 (b)
Grates for, of cast iron	31 (a)	28 (a)
Funeral wreaths of artificial flowers	350	292

	1905. Par.	1909. Par.
Fuses:		
For fireworks	111 (b)	91 (b)
For miners	111 (a)	91 (a)
Furniture of:		
Bamboo, rattan, wicker, cane, rushes, grass, or analogous materials	204	169 (b)
Bent wood	198	162
Other. (According to material and elaboration.)		
Furniture springs of wrought iron	45	47
Furs and fur skins		177
Gabanum, varnish	88	62
Galange root	81	57
Gallic acid	92 (d)	71 (b)
Gallnuts	81	57
Galloons. (According to material)		Rule 7, Sec. 5
Galvanic batteries	248	193
Galvanized iron or steel for roofing	36 (c)	31 (b)
Galvanometers	248	193 (a)
Gambling outfits, prohibited	Sec. 6	Sec. 6 (c)
Game	268	204
Games (toys), except those of gold, silver, etc	353	296
Gang plows	245	194
Garden seeds	302	227
Garnets in the rough or cut, unset	24, 379	13
Garters, elastic. (According to material.)		
Gaskets, rubber	352 (a)	293 (a)
Gas motors	243	194
Gasoline	10 (a)	23 (d)
Engines	243	194
Motors	243	194
Gauge liter, for tariff purposes, defined		Class XIII, Group 6
Gauges, steam pressure	257	194
Gauze:		
Bandages	99	82
Copper or brass	66 (a), (b)	49 (e)
Iron wire	44 (a), (b)	36 (d)
Other. (According to material.)		
Geese	268	175, 204
Gelatin	109	90
Capsules, empty, for medicine		81
Other manufactures of		90
Geldings		170, 342, 348
Gelgaum walnut oil	100 (b)	83
General Rules		Sec. 5
Generators, electric	250	193 (a)
Genista:		
Manufactures of	204	169 (c)
Unmanufactured	203	169 (a)
Gentian root	81	57
Geodetical appliances and apparatus, not otherwise provided for		303
Geranium, essence, oil	105 (a)	87 (a)
German silver, usually copper alloy. (See Copper.)		
Germatol	99	81
Germea	282	221
Ghee	336	275
Gimlets	46	194
Gin	308 (a)	263
Sloe	308 (c)	263
Essence of, concentrated	105 (a)	87 (a)
Gins, cotton	245	194
Ginger:		
Ale	312 (c)	268
Brandy	308 (b)	262
Grass, oil of	105 (a)	87 (a)
Ground	296 (b)	255 (b)
Unground	296 (a)	255 (a)

	1905. Par.	1909. Par.
Girths, saddle:		
Cotton	134	115
Linen, and other vegetable fibers	159	133
Ginseng root	81 (a)	58
Glass		Class 1
Articles—		
Not cut, enameled, engraved, gilt, or painted	13 (b)	16 (a),
		20 (b), (e)
Painted, enameled, engraved, gilt, or cut	13 (a)	16 (b), 20 (c)
Beads—		
Curtains of	16	20 (e)
Imitating amber	342 (b)	281
Loose, or in strings	16	20 (e)
In necklaces or other ornaments	340	279
Bottles—		
Common hollow glassware	12	15 (a)
Cigarette holders imitating amber	342 (b)	281 (b)
Cutters	46	196
Dry plates, enameled photographic	362	302
Graduates, engraved	13 (a)	20 (e)
Goblets	13 (a), (b)	16 (a), (b)
Hollow ware, ordinary	12	15 (a)
House decorations	13 (a), (b)	20 (b), (c)
Imitations of precious stones	26	14
Incandescent electric lamps	250 (a)	193 (a)
Insulators	248	193 (a)
Lamp:		
Chimneys	13 (c)	17
Globes	13 (b)	16
Mirrors	15	19
Paper	185	149
Paving slabs, cones, or prisms	14 (a)	18 (a)
Photographic dry plates	362	302
Plate	14	18 (c)
Enameled or engraved	14 (d)	18 (d)
Powdered or crushed	16 (b)	20 (d)
Roofing slabs, cones, or prisms	14 (a)	18 (a)
Spectacles	16 (a)	20 (a)
Sheets for windows (common)	14 (b)	18 (b)
Siphons for aerated water	12 (a)	15 (b)
Slides for magic lanterns	254	186
Statuettes, flower stands, etc	13 (a), (b)	20 (b), (c)
Stirring rods	16 (b)	20 (e)
Table service	13 (a), (b)	16 (a), (b)
Test tubes	16 (b)	20 (e)
Tinned (mirrors)	15	19
Toilet articles, urns	13 (a), (b)	20 (b), (c), (c)
Watch crystals	228 (a)	187
Wash basins	16 (b), (c)	16 (a), (b)
Glasses:		
For clocks	239	187
Looking	15 (a), (b), (c)	19
For spectacles	16 (a)	20 (a)
Reading	16 (c)	20 (e)
Glauber's salt	94 (c)	74 (c)
Glazed cardboard:		
In sheets	189 (a)	150 (a) 151 (a)
Manufactures of	190 (b), (d)	150 (b) 151 (b)
Globe valves, not machinery. (According to material.)		
Gloves:		
Baseball and boxing	220	179 (b)
Cotton, knit	125 (c)	107 (c)
Kid skin		179 (a)
Leather, other	220	179 (b)
Linen, knit	152 (c)	127 (b)
Silk	173 (b)	148
Wool	165 (b)	143
Other. (According to material.)		

	1905. Par.	1909. Par.
Grates, for furnaces, cast iron	31 (a)	28 (a)
Grease:		
Axle	8	22
Goose	101 (d), (e)	84
Graphite	89	67
Green soap, sapo viridis	104	86
Grindstones	2 (d)	3
Gross weight defined	Rule 17	Rule 13 (d)
Ground salt	94 (b)	74 (c)
Ground nut oil	100 (b)	83
Guajacol	99	81
Gualtheria oil	105 (a)	87 (a)
Guano	383	321
Guaracol	99	81
Guarana	81	57
Guayacum, extract of	99	81
Gulaman	109	90
Gum:		
Acacia, aloes, arabic, asafetida, benzoin, kino, sanda-rac, senegal, shellac, tragacanth, and other vegetable	78	56 (b)
Chewing	332	244
Colophony	77 (a)	56 (a)
Paper	187	150, 151
Gumabault metal. (See Copper.)		
Gumdrops	332	244
Gun:		
Cotton (pyroxylin, explosive)	111 (a)	91 (a)
Covers of leather	229	182
Powder. (See Sec. 6.)		
Explosive compounds, miners' fuses and caps	111 (a)	91 (a)
Other		91 (b)
Guns. (See Arms.)		
Gunny sacks	145	121
Gut, dried	233	183
Gutta-percha:		
Crude	77 (c)	293 (a)
Manufactures of	352	293
Gutters, iron	36	31 (b)
Gymnasium apparatus		297
Gypsum:		
Unwrought	3 (a)	6 (b), (c)
Wrought	4	7
Hacksaws	46	194
Haddock:		
Canned	317 (b)	215 (a)
Dried, in bulk	274 (b)	216 (b)
Fresh	274 (a)	216 (a)
Pickled or smoked, in bulk	274 (b)	216 (b)
Haematites		60
Hair:		
Animal, not further advanced than washed	161	314
Other, and manufactures of	164	139
Braids	164	139
Brushes. (According to material.)		
Clippers	54 (d)	44 (b)
Dyes	97, 99, 105 (b)	87 (b)
For violin bows, prepared	164	185
Human, manufactured or not	344	140
Manufactures of		Class VII
Mattresses	164	139
Pins, of common metals	52, 68	278
Other. (According to material.)		
Spun	163	169
Textiles	164	169
Tonic, perfumed	105 (b)	87 (b)

	1905. Par.	1909. Par.
Hair—Continued.		
Vegetable—		
Manufactured	204	169 (c)
Unmanufactured	203	169 (a)
Halter bits, saddlery hardware	49	42
Halters of leather	228 (c)	181 (b)
Hammers	46	194
Steam	257	194
Sledge	46	194
Hammocks:		
Of cotton		97
Of other vegetable fibers		120 (d)
Other. (According to material.)		
Hams:		
Canned or potted	315, 316	209
Deviled	315, 316	211
Smoked or cured, in bulk	270	207
Handkerchiefs. (According to material.)		
Making up defined, surtax for		Rule 11, Sec. 5
Handles:		
For tools and implements	194 (e)	194
For umbrellas. (According to material.)		
Hardware, saddlery, of wrought iron or steel	49 (a), (b)	42
Hard rubber manufactured in articles	352 (b)	293 (c)
Harika wares	191 (b)	156
Harmoniums	236	185
Harness:		
Draft, and parts of	228 (a)	181 (a)
Dressing	89	67
Finishings (see also Saddlery hardware)	49	42
Makers' wares, other	228 (c)	181 (b)
Other, and parts of		181 (b)
Harpoons	46	194
Harrows	245	194
Harvesting machinery	245	194
Hat:		
Bands, of sheepskin, finished	229	182
Boxes, of whatever material	228 (c)	288
Hatchets	46	194
Hats, and crowns for:		
Of paper		296
Of straw, chip, palm, grass, rattan, etc	355	299
Of other materials	356	300
Hat strings, manufactured of silk	174	148
Hay (forage)	303	228
Hayforks	46	194
Hazel, essence	105 (a)	87 (a)
Hazeline	99	81
Headings, hoops, bungs, staves, and shooks (coopers' wares)	194 (b)	160
Headwear, not specially mentioned		301
Hectogram, abbreviation for		Sec. 7
Hectoliter, abbreviation for		Sec. 7
Hedeoma, oil	99	87 (a)
Hedoral, carbolic disinfectant	97	77
Hedge shears	54 (c)	44 (a)
Heelball, shoemakers'	103	85 (c)
Helenin	99	81
Hellebori, extract	99	81
Hemlock oil	99	87 (a)
Hemp:		
Carpeting	155	129
Cinches	159	133
Cords, cordage, and rope makers' wares	144, 146	120
Corset laces	158	132
Export duties on		355
Fishing nets	144	120 (a)
Galloons	157	131

	1905. Par.	1909. Par.
Insulating materials, tapes, and compounds used exclusively for electrical purposes	248	193 (a)
Insulators, glass and porcelain	248	193 (a)
Integuments of animals		183
Intestines:		
Dried	233	183
In strings for musical instruments	233	185
Salted or in brine	269	206
Invoice, currency of		Sec. 18
Invoice forms, printed paper	179	151 (b)
Invoices:		
Consular	Sec. 22	Sec. 20
Frauds or intended frauds in, to be reported by American consular and customs officers to the insular collector of customs		Sec. 21
Requirements for		Secs. 18, 19, 20
Iodide of:		
Ammonia, arsenic, iron, strontium, and zinc	99	81
Potassium	97	77
Iodine	90 (b)	69
Iodoform	99	81
Ipecacuanha root	81	57
Iris root	81	57
Iridin	97	77
Iron:		
Carbonate of	94 (h)	74 (c)
Lactated	99	81
Citrate of	95 (b)	75
Iron and ammonium	99	81
Cocodylate hypophosphite, hypo and glycerophosphate, iodide, lactate, phosphate, potassic tartrate, and other pharmaceutical salts of	99	81
Oxide of (colors)	84 (c)	60, 61 (d)
Perchloride, solid, or solution of	97	77
Pure metallic, so-called reduced iron	99	81
Sulphate of	94 (c)	74 (c)
Iron, cast (if malleable, *see* Wrought iron)		Class III, Group 3
Articles not coated or ornamented with another metal or porcelain, neither polished nor turned	31 (a), (b), (c)	28 (a), (b)
Articles of all kinds, either polished or turned, enameled, bronze gilt, tinned, or coated; or with ornaments, borders, or parts of other metals (gold or silver excepted), or combined with glass or ceramic ware	32, 33	29
Bars and beams	31 (a)	28 (a)
Boxes, lubricating, for railway trucks and carriages	31 (b)	28 (b)
Chairs, for railways	31 (b)	28 (b)
Columns	31 (a)	28 (a)
Cuttings and filings, fit only for resmelting	74	309
Grates for furnaces	31 (a)	28 (a)
Lamps of	32, 33	29
Plates	31 (a)	28 (a)
Pipes	31 (a)	28 (a)
Fittings for	31, 32, 33	28 (b), 29
Pigs	30	309
Wastes and shavings, fit only for resmelting	74	309
Iron or steel, wrought		Class III, Group 3
Anchors	42	310
Anvils	42	38
Articles not specially mentioned	58, 59	47
Axles—		
For railways and tramways	38 (a)	33 (a)
For vehicles	38 (b)	33 (b)

	1905. Par.	1909. Par.
Iron or steel, wrought—Continued.		
Bamboo steel in bars	35 (b)	30 (c)
Barbed wire (fencing)	45	36 (a)
Barrels for firearms	55 (a), (b)	45
Bars of—		
Crucible steel	35 (c)	30 (d)
Wrought iron or steel	35 (b)	30 (c)
Beams, wrought iron or steel	35 (b)	30, 34
Bolts	47	39
Brads	50 (a), (b)	43
Buckles	50 (a), (b)	43
Buttons	345 (a)	283 (b)
Cables	45	36 (a)
Cast, in pieces in the rough	37 (a), (b)	32
Chairs of, polished	59	47
Railway		30 (a), (b)
Chains	42	37
Clasp nails	48	40
Cold rolled, corrugated, or galvanized sheets	36 (c)	31 (b)
Crochet hooks	53	278
Cranks	38 (b)	32, 33, 194
Cutlery	54	44
Dental instruments	54 (e)	196
Fencing (barbed wire)	45	36 (a)
Firearms and detached parts for	55, 56	45
Fish plates	38 (a)	30 (a), (b)
Fishhooks	54 (d)	44 (a)
Furniture springs	45	47
Gauze, wire	44 (a), (b)	36 (d)
Hairpins	53	278
Hooks and eyes	52	278
Hoops	35 (b)	31 (b)
Hoop iron and fledges	36 (a)	31 (b)
Horseshoes	58	47
Ingots, or "tochos"	34	309
Instruments, surgical or dental	54 (e)	196
Lubricating boxes	38 (a)	33 (a)
Mooring buoys	42	47 (a)
Nails	48	40
Needles (except surgical)	51	278
Surgical		196
Netting	45	36 (b)
Nuts	47	39
Pens	51	278
Perforated sheets	36 (c)	31 (b)
Pieces, large, composed of bars, or bars and sheets fastened together or not (structural iron)	41	34
Pins, common and safety	52	278
Pipes	39 (a), (b)	35
Fittings for	58, 59	47
Pistols	55 (c)	45
Plates	42	31 (a)
Pocket cutlery	54 (b)	44 (b)
Polished sheets	36 (c)	31 (b)
Rails	35 (a)	30 (a), (b)
Revolvers, and detached parts for	55 (c)	45
Razors	54 (b)	44 (b)
Rivets	47	39
Rods	35 (b), (c)	30 (c), (d)
Saddler's hardware	49 (a), (b)	42
Safes	58, 59	47
Scissors	54 (a), (b), (d)	44 (a), (b)
Screws	47	41
Sheets, rolled	36 (a), (b), (c)	31 (a), (b)
Side arms (not fire), and parts for	54 (b)	44 (b)
Sleepers	38 (a)	30 (a), (b)
Spirit levels	46	194

	1905. Par.	1909. Par.
Iron or steel, wrought— Continued.		
Springs—		
For railways and tramways	38 (a)	33 (a)
For vehicles	38 (b)	33 (b)
Furniture	45	47
Structural steel, cut to measure	41	34
Surgeons' operating tables	59	47
Instruments		196
Switches, rails, and tongues (*see* note)	42, 35 (a)	30 (a), (b)
Tacks	48	41
Tires	35 (b)	30 (c)
Tinned sheets, terne and tin plate	36 (b)	31 (c)
Manufactures of	57	46
Tools and implements	46	194, 196
Trunks	58, 59	288
Umbrella frames	59	47, 298 (d)
Washers	47	39
Watch chains, adornments	340	279
Wheels	38 (a), (b)	33 (a), (b)
Wire—		
Galvanized or not	40, (a), (b), (c)	36
Gauze	44 (a), (b)	36 (d)
Netting	45	36 (b)
Irrigation pumps	257	194
Isinglass, gulaman, gelatin, and manufactures thereof	109	90
Ivory:		
Black	89	61 (c), (d)
Combs	339 (b)	280 (b)
Compositions imitating	342	281
Fans	338	277
Unwrought	341 (a)	280 (a)
Wrought	341 (b)	280 (b)
Jack screws, lifting jacks, hoisting apparatus	245	194
Jade		13
Jams, fruit	322 (b)	237
Japalac (varnish)	88	62
Japanese:		
Lanterns, paper	190 (d)	151 (b)
Napkins, paper	190 (d)	150, 151
Saki (wine)	310	265, 266
Soy	328	256
Toyo	328	256
Jardinieres:		
Of ceramic wares		11 (f)
Other. (According to material.)		
Jasmine oil	105 (a)	87 (a)
Jasper	1	1
Jelly, fruit	332	237
Jerked beef	269	207
Jerseys. (According to material.)		
Jet:		
Compositions imitating	342	281
Unwrought	341 (a)	280 (a)
Wrought	341 (b)	280 (b)
Jewelry:		
Broken up, of gold, silver, and platinum	372	318
Gold platinum, or alloys thereof, not otherwise provided for	27 (a)	25 (a)
Set with pearls or precious stones	27 (b)	25 (b)
Set with doublets or imitation precious stones or pearls	27 (c)	25 (c)
Gold plated	29 (a)	27 (a)
Silver—		
Not set with pearls or precious stones or imitation precious stones or imitation pearls	28 (a)	26 (a)
Set with pearls or precious stones	28 (b)	26 (b)
Set with imitation pearls or imitation precious stones	28 (c)	26 (c)
Plated	29 (a)	27 (a)

Jewelery—Continued.

	1905. Par.	1909. Par.
Unfinished parts for rings and earrings, and similar nufinished portions of—		
Gold	27 (d)	25 (d)
Silver	28 (d)	26 (d)
Gold or silver plated	29 (b)	27 (c)
Jewels (*see* Precious stones)	24	13
Used, imported by passengers in their baggage	386	339, 341
Jinrikshas	261	201
Jiri, Indian spices	296	255 (b)
Jodelite (unrefined coal tar)	7	22
Jonquil oil	105 (a)	87 (a)
Joss:		
Money	190 (d)	151
Sticks	105 (b)	87 (b)
Juice of fruit, unfermented	313	268
Other		269
Juniper:		
Oil, essential	99	87 (a)
Tar	99	81
Jute		Class VI
Carpeting	155	129
Cinches	159	133
Cordage and rope-makers' wares	146 (a), (b)	120
Corset laces	158	132
Fibers, raw	138	313
Galloons	157	131
Gunny bags	145	121
Knitted goods	152	127
Laces and tulles	154	128
Laces for corsets and shoes	158	132
Manufactures of	136 to 160	Class VI
Plushes, velvets, velveteens, and pile fabrics	151	126
Raw	138	313
Ribbons	157	131
Ribbons or bands for the manufacture of cinches and saddle girths	159 (a)	134
Rope	146	120
Saddle girths	159	133
Shoe laces	158	132
Tape	157	131 (a)
Textiles	147, 148, 149, 150	Class VI, Group 2
Textiles called tapestry	156	130
Threads, twines, cords, and yarns, netted hammocks, fishing nets, and similar manufactured articles	144	120
Trimmings	157	131
Tulles	153	128
Waterproof or caoutchouc stuffs on textiles of	160	135
Yarns	140, 141	119, 120
Kakamonos of straw	204 (a)	196 (c)
Kal, carbonic, puris	97	77
Kalium, ferric tartaric	99	81
Kananga face powder, and water	105 (b)	87 (b)
Kangaroo tendons:		
Dried	82	183
Aseptic, for surgical use	99	82
Kangaroos	211	174
Kaolin	3 (b)	6 (b), (c)
Karlsbad salts	99	81
Kekume oil	100 (b)	83
Kettles, cast-iron	31 (a), 31 (c)	28 (b), 29
Khaki blanco for shoes	89	67
Kids	211	174
Kid skin:		
Curried	217 (c)	178 (b), (c)
Gloves	220	179 (a)
Shoes of	223	180 (b)

	1905. Par.	1909. Par.
Kieselguhr (earth)	3 (b)	6 (b), (c)
Kilogram equals 2.2046 pounds avoirdupois	Sec. 9	Sec. 10
Abbreviation for		Sec. 7
Kinetoscope and parts, appurtenances, and accessories for.		186
Kino gum	78	56 (b)
Kitchen:		
Knives	54 (a)	44 (a)
Utensils of stoneware	19 (a)	11 (b), (c)
Kits, film		302
Knife switches, electrical	248	193 (a)
Knitted goods. (According to material.)		
Admixtures of materials in		Rule 6, Sec. 5
Knives:		
Bread, kitchen, plumbers', painters', saddlers', butchers', shoemakers', and cheese	54 (a)	44 (a)
Corn and hemp	46	194
Hunting		44 (b)
Pocket	54 (b)	44 (b)
Pruning and budding	54 (c)	44 (a)
Sheath		44 (b)
Table—		
Wood or iron handles	54 (a)	44 (a)
Other handles, common	54 (d)	44 (b)
Of silver	28 (a), (d)	26 (a), (d)
Of silver plate	29 (b)	27 (b)
Other cutlery	54 (d)	44 (b)
Kola:		
Astier	99	81
Extract	99	81
Nut	81	57
Komel	312 (c)	268
Kukui oil	100 (b)	83
Kukum (Indian vegetable product)	81	57
Kummel	308 (c)	263
Labels:		
Plain		150 (b)
Printed	179	151 (b)
Lithographed	183	151 (b)
Lace paper	190 (d)	151 (b)
Laces. (According to material.)		
Admixtures of materials in		Rule 6, Sec. 5
Shoe. (According to material.)		
Lactate:		
Iron	99	81
Of calcium	99	81
Lactopeptin	99	81
Lactophenin	99	81
Lactophosphate calcii	99	81
Lactocarium	99	81
Ladano, gum resin	78	56 (b)
Lag bolts	47	39
Lambs	211	174
Lambskin:		
Curried	217 (c)	178 (b), (c)
Shoes	223	180 (b)
Lamp:		
Black	89	61 (c), (d)
Chimneys, glass	13 (c)	17
Globes, glass	13 (b)	16 (a), (b)
Wicks—		
Cotton	130	112
Vegetable pith (timsim)	366	169 (a)
Lamps:		
Arc	248	193 (a)
Bicycle	252	199
Of brass	69 (a), (b)	50

	1905. Par.	1909. Par.
Lamps—Continued.		
Of glass	13	16
Glass chimneys for	13 (c)	17
Incandescent	250 (a)	193 (a)
Wicks, cotton for	130	112
Other. (According to material.)		
Lanolin (wool fat)	101 (a), (b)	84
Lanoline, perfumed for toilet use	105 (b)	87 (b)
Lanterns, magic, parts and accessories for		186
Lanterns, paper	190 (d)	151 (b)
Other. (According to material.)		
Lap robes. (According to material.)		
Lacquered canvas shoes, imitating patent leather	222	180 (b)
Lard, and imitations of	271, 272	208
Lard oil	101 (d), (e)	84
Lasts, common wood	195	162
Lathes	257 (b)	194
Laths, common wood	192 (a)	157 (b)
Latticework, wood	194 (d)	162, 163, 164
Laughing gas	97	77
Launches	267	203
Laurel:		
Essence (oil of)	99	87 (a)
Leaves	81	57
Lava, and manufactures of		5
Lavandula spica, oil of	99	87 (a)
Lavender flowers, oil of	105 (a)	87 (a)
Lawn mowers	257 (b)	194
Lead:		
Acetate of (sugar of lead)	95 (a)	75
Carbonate and oxide of	84	61 (a)
Lead, and alloys of:		
Bars	73 (b)	54 (a)
Bullets		54 (b)
In lumps or ingots, pure	73 (a)	309
Alloys of		54 (a)
In nickeled or bronze gilt articles	73 (c)	54 (c)
In articles not nickeled or bronze gilt	73 (f)	54 (b)
Pipes	73 (b)	54 (a)
Sheets	73 (b)	54 (a)
Shot and bullets	73 (b)	54 (b)
Solder, in bars	73 (b)	54 (a)
Traps, plain, for sanitary construction		54 (a)
Type	73 (f)	54 (a)
Wire	73 (b)	54 (a)
Lead, red and white	84 (a), (b)	61 (a), (b)
Lead pencils	85 (c)	65
Leaf, gold	27 (d)	25 (d)
Leather:		
Belting for machinery	229 (a)	195
Boots and shoe findings		178 (a), (b), (c)
Calfskin, curried	217 (b)	178 (b), (c)
Chamois	219	178 (c)
Gloves	220	179
Hat boxes	228 (c)	288
Hides and skins	214, 215, 216	176, 177, 178
Pillows, covered with	229	182 (b)
Saddlery and harness makers' wares	228	181
Shawl straps	229	288
Sheepskins, curried (basils)	217 (a)	178 (a)
Shoes of—		
Calfskin	223	180 (b)
Canvas lacquered, imitating patent leather	222	180 (b)
Cow and horsehide	221	180 (a)
Goatskin	223	180 (b)
Kid	223	180 (b)
Lambskin	223	180 (b)
Sheepskin	221, 223	180 (a)

	1905. Par.	1909. Par.
Leather—Continued.		
Other manufactures of	229	182
Trunks, valises, traveling bags, and similar receptacles for personal effects	228 (c)	288
Leaves, as drugs, such as buchu, erysimum, laurel, sage, thyme, rosemary, and senna	81	57
Sage or thyme, ground for culinary purposes	296	255 (b)
Leaves, artificial, and parts of		292
Lecithine	99	81
Tabloids	98	81
Ledger paper	179	150, 151
Leeches	365	174
Lees, wine		76
Lemon:		
Oil	105 (a)	87 (a)
Salts, effervescent	99	81
Lemon grass oil	105 (a)	87 (a)
Lenses:		
For cameras	361, 362	302
For spectacles or eyeglasses	16 (a)	20 (a)
Other	16 (c)	20 (e)
Leopards	211	174
Letter:		
Balance scales	242	192
Balance scales	242	192, 151 (b)
Paper	179	150, 151
Press copy book	187	153
Library paste	109	90
Lichees:		
Canned	322 (b)	236
Dried	286	235
Fresh	374	234
Lichens	81	57
Licopodium or lycopodium	81	57
Licorice:		
Extract of	78	81
Root	81	57
Lifeboats	267	203
Life buoys		310
Life preservers		310
Lifting jacks	245	194
Ligatures, catgut, silk, and the like, for surgical use, aseptic	99	82
Lighters, cigars, electric		193 (b)
Lighters, paper	190 (d)	150 (d)
Lighters (boat)	267	203
Lightning arresters	248	193 (a)
Lights, developing		302
Lilac oil	105 (a)	87 (a)
Lily:		
Flowers, dried (Chinese)	288	231
Root flour	282	221
Nuts	366	239
Limbs and members, artificial		282
Limburger cheese	334 (b)	273
Lime:		
Chloride of (calcium hypochlorite)	94 (e)	74 (b)
Hydraulic	3 (a)	6 (a)
Phosphate and superphosphates of	94 (d)	74 (a)
Stone	2	2
Linden flowers	81	57
Linen		Class VI
Cinches	159	133
Gloves	152 (c)	127 (b)
Knitted goods	152	127
Plushes, velvets, pile fabrics	151	126
Saddle girths	159	133

	1905. Par.	1909. Par.
Linen—Continued.		
Shoe and corset laces	158	132
Socks and stockings	152 (c)	127 (a)
Textiles		Class VI, Group 2
Trimmings	157	131
Tulles and laces	153	128
Liniments	99	81
Lini oil	100 (b)	83
Linoleum	349 (d)	287
Linseed:		
Crushed, in meal or in cake	81	55 (b)
Whole	76	55 (a)
Oil	100 (b)	83
Lions	211	174
Liquors	308 (c)	263
Liquors, adulterated, prohibited		Sec. 6 (e)
Alcoholic	308	261, 262, 263
Malt	312 (a), (b)	267
Vinous	309, 311	264, 265, 266
Listerine	99	81
Liter (equivalent to 1.0567 quarts, wine measure)	Sec. 9	Sec. 10
Liter, gauge and proof, for tariff purposes, defined		Class XIII, Group 6
Litharge, color	84 (c)	61 (c), (d)
Lithium, carbonic	97	77
Lithographic:		
Inks	85 (a)	64 (a)
Presses	257 (b)	194
Stones	1	1
Lithopgraphs	183	151
For advertising, when free	381	327
Litmus, color	85	61 (c), (d)
Litmus, paper	187	151
Live animals	205, 211	170–175
Loaded dice (prohibited)	Sec. 6	Sec. 6 (c)
Locomotives and tenders	246	194
Logs:		
Fine wood	•193 (a)	158 (a)
Ordinary wood	192 (a)	157 (a)
Logs, ships' (instruments)	257 (a)	303
Logwood extract	87 (a)	66 (b)
Loofah sponges	365	285
Looking glasses	15	19
Lottery tickets; advertisements of, or lists of drawings, prohibited		Sec. 6 (f)
Lubricating:		
Boxes—		
Of cast iron	31 (b)	28 (b)
Of wrought iron or steel	38 (a)	33 (a)
Oils—		
Mixed, crude	8	22
Refined	9	23 (b)
Axle greases of all kinds	8	22
Lucca oil	304	83
Lucky paper	187	150, 151
Lumber	192, 193	157, 158
Lupulin	81	57
Lychees:		
Dried	286	235
Fresh	374	234
Canned	320	236
Lycopodium	81	57
Lye, caustic	93	72
Lysol	97	77
Lysolum	97	77
Macaroni	284	225

	1905. Par.	1909. Par.
Maccasar, oil of	105 (a)	87 (a)
Mace:		
Ground	297 (b)	250 (b)
Unground	297 (a)	250 (a)
Oil of	105 (a)	87 (a)
Machetes	46	194
Machine needles, sewing	51	278
Machinery and apparatus		Class XII, Group 2
Agricultural	245	194
Boilers	244	194
Compressed-air engines and motors	243	194
Cranes, hand or power	247	194
Derricks	245	194
Detached parts of machines are dutiable under the corresponding paragraphs under which the complete machine is assessed.		
Drill grinder, twisted	257 (b)	194
Electric, of all kinds		193
Engines, of all kinds	243	194
For dredging, hoisting, or pile driving	245	194
For extracting oil from cocoanuts	245	194
Gasoline motors	243	194
Hand pumps	257	194
Hemp stripping	245	194
Hot-air motors	243	194
Hullers, cleaners, threshers, rice	245	194
Hydraulic cranes	247	194
Ice-making	245	194
Jackscrews, hoisting	245	194
Locomotives	246	194
Meters—		
Electrical, of all kinds	248	193 (a)
Other		187, 194
Motors—		
Hydraulic, steam, petroleum, gasoline, and hot or compressed air	243	194
Electrical, all kinds	250	193 (a)
Pile-driving	245	194
Pumps, hand or power	257	194
Refrigerating	245	194
Rice-hulling	245	194
Road-making	245	194
Sawmill	245	194
Stamp mills	257 (b)	194
Steam boilers of all kinds	244	194
Steam plows	245	194
Steam pumps	257	194
Sugar mills	245	194 (a)
Tenders to locomotives	246	194 (b)
Turbine engines	243	194
Turntables	247	194
Typewriters	253	188
Weighing	242	192
Windlasses	245	194
Windmills	257 (b)	194
Wood working	257	194
Machines. (See also Machinery.)		
Adding	255	188
Aerial and parts for		201
Automatic slot, if not prohibited	256	186, 191
For gambling, prohibited	Sec. 6	Sec. 6
Boring	257 (b)	194
Cash registers	255	189
Cigarette making	257 (b)	194
Corking	257 (b)	194

	1905. Par.	1909. Par.
Machines—Continued.		
Cotton gin	245	194
Duplicating		188
For drying tobacco	257 (b)	194
Hammers, steam	257 (b)	194
Ice-cream freezers	257 (b)	194
Kinetoscopes	254	186
Manifolding		188
Phonographs	254	186
Planing	257 (b)	194
Sewing	251	190
Shears	257 (b)	194
Washing	257 (b)	194
Water meters	257 (a). (b)	194
Writing		188
Mackerel:		
Canned or potted	317 (b)	215 (a)
Fresh	274 (a)	216 (a)
Salted, smoked, or pickled, in bulk	74 (b)	216 (b)
Mackintoshes. (According to material.)		
Madder	86 (b)	66 (a), (b)
Magazines	382 (b)	328
Magic lanterns and parts therefor	254	186
Magnesium, metallic, in ribbons	73 (f)	54 (b)
Magnesium:		
Calcined oxide	99	77
Carbonate	94 (c)	74 (c)
Chloride	94 (h)	74 (c)
Citrate—		
Effervescent	99	81
Other	95 (b)	75
Glycerophosphate	99	81
Metallic, in powder or ribbons	73 (f)	54 (b)
Nitrate	94 (h)	74 (c)
Phosphate	94 (h)	74 (c)
Salicylate	99	81
Sulphate	94 (c)	74 (c)
Magnets, horseshoe	46	194
Magnifying glasses	16 (c)	20 (e)
Maize	278	220
Oil of	100 (b)	83
Making up defined, surtax for		Rule 11, Sec. 5
Malleable cast iron. (Dutiable as wrought iron.)		
Mallets	195, 196	194
Polo		297
Malt	280	319
Beverages	312	267
Extracts, pharmaceutical	99	81
Whisky	308 (a)	263
Roasted (a coffee substitute)	282	221
Malted Milk	282	222
Manganese:		
Black oxide	83	60
Borate	99	81
Glycerophosphate	99	81
Peptonized	99	81
Pepotonized	99	81
Manganesit (putty)	84 (b)	61 (b)
Mangan putty	84 (b)	61 (b)
Manifolding machines and parts for		188
Manila paper	178	150, 151
Manna	78	56 (b)
Mannit	99	81
Mantillas. (According to material.)		
Mantels, Welsbach	365	5
Manufactures of the Philippine Islands, reimported		350

	1905. Par.	1909. Par.
Manures, natural	383	321
Manuscripts, free		329
Maple sirup	337	241
Maps	182	155
Maraschino cherries	323	238
Maravilosa essence	99	81
Marble	1	1
Dust	3 (b)	1 (a)
In rough	1 (a)	1 (a)
Other articles of	1 (d)	1 (c)
Plates	1 (b)	1 (b)
Sculptures	1 (c)	1 (c)
Slabs	1 (b)	1 (b)
Steps	1 (b)	1 (b)
Marbles (toys)	353	296
Mares	205 (b)	170, 333, 348
Marine engines	243	194
Marjoram oil	99	87 (a)
Marked playing cards (prohibited)	Sec. 6	Sec. 6 (c)
Market value, defined		Rule 13 (a), Sec. 5
Marking ink	85	64 (b)
Marmalade	322 (b)	237
Masks for baseball or fencing of whatever material		297
Masks, face		296
Massey oil	105 (a)	87 (a)
Mastic	78	56 (b)
Matches of all kinds	351	92
Materials:		
Component, defined		Rule 12, Sec. 5
Insulating	248	193 (a)
Not otherwise provided for	366	306
Mathematical appliances and apparatus not otherwise provided for		303
Matricaria chamomilla (German chamomile)	81	57
Mats. (According to material.)		
Matte, copper, free		309
Mattocks	46	194
Mattolein	97	77
Mattresses, for beds. (According to material.)		
Maypole soap	87 (b)	66 (c)
Maximum rate of duty on importations		Sec. 11
Measuring tapes:		
Steel	46	194
Other. (According to material.)		
Meat:		
Canned	315, 316	209, 210, 211
Extracts		213
Fresh	376	205
Juices		213
Salted or in brine	269	206
Smoked or cured	270	207
Soup tablets		213
Mechanical music boxes	237	185, 186
Medals, as trinkets or ornaments	27, 28, 340	25, 26, 27, 279
As trophies or prizes	27, 28, 340	330
Other. (According to material.)		
Medical induction batteries	54 (e)	193 (b)
Medicated:		
Cotton	99	82
Soap	104	86
Medicines, adulterated or deleterious, prohibited		Sec. 6 (e)
Medicines, patent or proprietary	99	80
Other		81
Meerschaum:		
Unwrought	341 (a)	280 (a)
Wrought	341 (b)	280 (b)

	1905. Par.	1909. Par.
Melaleuca flaviflora oil	99	87 (a)
Mellisa oil	105 (a)	87 (a)
Mellots metal. (*See* Zinc.)		
Menthæ piperitæ oil	105 (a)	87 (a)
Menthæ viridis oil (oil of spearmint)	99	87 (a)
Menthol	99	81
Menthol cones	99	81
Merchandise, not otherwise provided for	365	306
Mercerized cotton yarns	116	95
Mercury	70	51
Ammoniacal (white precipitate)	97	77
Bijodido (bi-iodide)	99	81
Chloride	94 (h)	74 (c)
Cyanide	94 (h)	74 (c)
Ointment	99	81
Oxide	97	77
Oxinate	97	77
Pills (blue mass)	98	81
Powdered mass	99	81
Merluza guisada, in cans	318	215 (c)
Metal:		
Bell, free		309
Muntz		309
Threads, in textiles	Rule 11	Rule 10
Or spangles, contained in embroidery		Rule 9, Sec. 5
Wastes, common, fit only for resmelting	74	309
Metallic threads, textiles of		Rule 10, Sec.5
Metals, not otherwise provided for, and alloys thereof	73	54
Meter, equivalent in inches (39.37 inches)	Sec. 9	Sec. 10
Meters:		
Electric, of all kinds	248	193 (a)
Water	257	194
Methyl:		
Acetate	97	77
Alcohol	307	260
Salicylas, oil	105 (a)	87 (a)
Methylal	99	81
Methylene blue	87 (b)	66 (c), 81
Methylene violet (medicinal), or pyoktanin	99	81
Metilal	97	77
Metol	' 97	77
Metrical system	Sec. 9	Sec. 10
Metronomes		185
Mexican lignaloes oil	105 (a)	87 (a)
Miaouli oil	99	87 (a)
Mica and manufactures thereof	2	5
Microscopes and slides for	365	303
Microtomes		303
Migranin	99	81
Milks and creams:		
Compounded with other substances (except sugar)		271
Condensed, concentrated, or evaporated	320	270
Fresh	378	270
Malted	282	222
Powder		271
Sugar of	99	81
Tablets		271
Millboards of asbestos	2 (e)	4
Millefleurs, extract for making perfumes	105 (a)	87 (a)
Millet:		
Flour	279 (b)	220 (b)
In grain	279 (a)	220 (a)
Mills of all kinds	245, 257 (b)	194
Millstones	2 (d)	3
Mimeograph ink	85	64 (b)

	1905. Par.	1909. Par.
Muria puama:		
Extract	99	81
Wood	81	57
Muriate of:		
Gold, silver, and platinum	96	78
Tin	94 (h)	74 (c)
Muriatic acid	91 (a)	70 (a)
Music:		
Boxes	237	185, 186
In raised print, used exclusively by the blind	390 (b)	328
Printed or lithographed sheet	180 (a)	151
Musical instruments	195, 197, 233, 235, 236, 237, 353	185, 186
Parts of, appurtenances and accessories for		185
Musk	105 (a)	87 (b)
Mustard:		
Dressing	328	256
Ground	300 (c)	253 (c)
In paste	300 (c)	253 (c)
Oil—		
Black	99	83
Volatile	99	87 (a)
Plaster	99	81
Seed	300 (a)	253 (a)
Mutton (see Meats, canned)	315, 316	209
Myrbane oil	97	77, 263
Myrciae oil	99	87 (a)
Myristicae oil, essential	105 (a)	87 (a)
Myrrh	81	57
Oil of	99	87 (a)
Myrtle oil	105 (a)	87 (a)
Nails:		
Of copper	68 (a), (b)	50
Of iron	48	40
Of zinc	73 (e)	54 (b)
Naphtha	10	23 (d)
Naphthalin	97	77
Naphthaline balls	97	77
Naphthol	99	81
Naphthol benzoin	99	81
Napkins. (According to material.)		
Making up, defined, surtax for		Rule 11, Sec. 5
Nasturtium oil	105 (a)	87 (a)
National issue coins and currency, free		320
Natrium. (See Sodium.)		
Neatsfoot oil	101 (d), (c)	84
Neckties. (According to material.)		
Needles:		
Of bone	342 (b)	281 (b)
Of common metals	53, 68 (c)	278
Of ivory	341 (b)	280 (b)
Phonograph	254	186
Sewing	51	278
Sewing machine	51	278
Surgical	54 (e)	196
Neroli oil	105 (a)	87 (a)
Nets. (According to material.)		
Netted and knitted stuffs, admixtures of materials in		Rule 6, Sec. 5
Netting, iron wire	45	36 (b)
Net weight defined	Rule 18	Rule 13
Newspapers, when free	382 (b)	328
Old, imported for wrapping paper	178	151 (a)
Nickel:		
Ammonium sulphate	94 (h)	74 (c)
Carbonate of	94 (h)	74 (c)
Sulphate of	94 (h)	74 (c)

	1905. Par.	1909. Par.
Nickel and alloys of:		
In bars, sheets, pipes, and wires...................	71 (b)	52 (a)
In lumps and ingots, alloys......................	71 (a)	52 (a)
Pure..		309
In other articles...............................	71 (c)	52 (b)
Nigrosine..	87 (b)	66 (c)
Nippers (pliers).................................	46	194
Nipples, rubber.................................	352 (d)	293 (b)
Nitrate of:		
Copper......................................	94 (h)	74 (c)
Potassium...................................	94 (d)	74 (a)
Silver.......................................	94 (h)	74 (c)
Silver in pencils for cauterizing...............	99	81
Sodium......................................	94 (d)	74 (a)
Strontium...................................	94 (h)	74 (c)
Uranium.....................................	99	81
Nitric acid.....................................	91 (a)	70 (a)
Nitroglycerin...................................	111 (a)	91 (a)
Note paper.....................................	179	150, 151
Novels...	180	154
Number of threads in textiles....................	Rule 1	Rule 1, Sec. 5
Nursing bottles, glass...........................	16	15 (a), 20 (e)
Nut galls.......................................	81	57
Nutmeg oil, essential............................	105 (a)	87 (a)
Nutmegs:		
Ground......................................	296 (b)	251 (c)
Husked......................................	298 (b)	251 (b)
Unhusked....................................	298 (a)	251 (a)
Nuts, edible....................................	329, 365	239
Cocoanuts...................................	76	55 (a)
Conserved or crystallized....................	331	238
For drugs....................................	81	57
Of copper....................................	68 (a), (b)	50
Of iron......................................	47	39
Products.....................................		239
Nux vomica.....................................	81	57
Tincture and medical preparations of..........	99	81
Okum..	365	311
Oars of wood...................................	195	162
Oatmeal, prepared for table use.................	282	221
Oats in:		
Grain.......................................	278 (a)	219 (a)
Meal or flour................................	278 (b)	219 (b)
Obscene, indecent, or subversive of public order, books, printed, or written matter, objects, prohibited.........		Sec. 6 (b)
Ochers, natural colors...........................	83	60, 61 (d)
Odometers and parts therefor....................		187
Officers and employees of the United States and insular go ernments to be considered as "settlers"..........		341
Oils v		
Aleuritis tribola (Kekume oil, Gelgaum walnut, Kukui oil......................................	100 (b)	83
Allspice (Ol. pimentae)........................	99	87 (a)
Almond, expressed............................	99	83
Amber (Ol. Succine)..........................	99	87 (a)
American wormseed (Ol. Chenepodii)............	99	87 (a)
Animal.......................................	101	84
Anise (Ol. Anisi).............................	105 (a)	87 (a)
Apricots.....................................	105 (a)	87 (a)
Arachidis....................................	100 (b)	83
Arachis (ground nut oil)......................	100 (b)	83
Asafetida....................................	99	87 (a)
Asarum......................................	99	87 (a)
Balm..	105 (a)	87 (a)
Banana, artificial............................	97	87 (a), 263
Bay (Ol. Myrcias)............................	105 (a)	87 (a)
Betula (see Birch)	105 (a)	87 (a)

Oils—Continued.	1905. Par.	1909. Par.
Ben	99, 100 (b)	83
Bengoin	105 (a)	87 (a)
Bergamot (Ol. Bergamotte)	105 (a)	87 (a)
Birch	105 (a)	87 (a)
Bitter almond (Ol. Amygdale amarae)	105 (a)	87 (a)
Black mustard	99	83
Volatile		87 (a)
Black pepper	99	83
Volatile		87 (a)
Blesemi	100 (b)	83
Cade (Ol. Cadinum)	99	87 (a)
Cajuput (Ol. Cajuputi)	100 (b)	87 (a)
Calamus (sweet flag)	105 (a)	87 (a)
Camphor	99	87 (a)
Canada fleabane (Erigontis)	99	87 (a)
Caraway (Ol. Cari)	99	87 (a)
Carob	99	87 (a)
Cassia (see Cinnamon)	105 (a)	87 (a)
Castor	100 (b)	83
Cedar Wood (yellow)	99	87 (a)
Celery	99	87 (a)
Ceylon cinnamon (Ol. Cinn. Zeylanci)	105 (a)	87 (a)
Chamomile (Ol. Anthemidis	99	87 (a)
Chenopodium	99	87 (a)
Cinnamon (Ol. Cinnamoni)	105 (a)	87 (a)
Citronella	105 (a)	87 (a)
Cloves (Ol. Caryophyli)	105 (a)	87 (a)
Cocoanut	100 (a)	83
Cod liver	99, 101	84
Prepared as pharmaceutical products	99	81
Cognac	105 (a)	87 (a)
Copaiba (Ol. Copaibae)	99	87 (a)
Coriander (Ol. Coriandri)	99	87 (a)
Cotton seed (Gossypii seminis)	100 (b)	83
Croton (Ol. Tiglii)	99	83
Cubeb (Ol. Cubebae)	99	87 (a)
Cumin	105 (a)	87 (a) .
Dill (Ol. Anethi)	105 (a)	87 (a)
Ergot	99	83
Erigeron (see Fleabane)	99	87 (a)
Essential	105 (a)	87 (a)
Eucalyptus (Ol. Eucalypti)	99	87 (a)
Euphorbia	99	87 (a)
Fennel (Ol. Foeniculi)	105 (a)	87 (a)
Flaxseed	100 (b)	83
Fleabane (Ol. Erigerontis)	99	87 (a)
Foeniculi	105 (a)	87 (a)
Fruit, artificial	308 (c), 97	77, 263
Gaultheria	105 (a)	87 (a)
Gelgaum	100 (b)	83
Geranium	105 (a)	87 (a)
Gin	105 (a)	87 (a)
Ginger	99	87 (a)
Ginger grass	105 (a)	87 (a)
Gossipii (cotton seed)	105 (a)	83
Hazel	105 (a)	83
Hedeoma (Ol. Hedeomae)	99	87 (a)
Hemlock (spruce oil)	99	87 (a)
Horseradish	99	87 (a)
Illicum religiosum (Shikimi: same as Indian anise)	105 (a)	87 (a)
Indian corn	100 (b)	83
Jasmine	105 (a)	87 (a)
Jonquil	105 (a)	87 (a)
Juniper oil (Ol. Juniperi)	99	87 (a)
Juniper tar	99	81

Oils—Continued.	1905. Par.	1909. Par.
Lard oil	101 (d), (e)	84
Laurel	99	87 (a)
Lavender flower	105 (a)	87 (a)
Lemon (Ol. Lemonis)	105 (a)	87 (a)
Lemon grass (Ol. Andropogon citrati)	105 (a)	87 (a)
Linseed (Ol. Lini)	100 (b)	83
Lilac	105 (a)	87 (a)
Lillies	105 (a)	87 (a)
Lucca	304	83
Mace	105 (a)	87 (a)
Maccasar	105 (a)	87 (a)
Maize	100 (b)	83
Marjoram	99	87 (a)
Massey (see Cinnamon)	105 (a)	87 (a)
Matricaria Chamomilla	99	87 (a)
Melaleuca flaviflore (Miaouli)	99	87 (a)
Mellisa	105 (a)	87 (a)
Mexican lignaloes	105 (a)	87 (a)
Mignonette	105 (a)	87 (a)
Mineral	9, 10	22, 23
Mirbane	97	77, 263
Mustard, volatile (Ol. Sinapis volatile)	99	87 (a)
Myrcia (see Oil of bay)	105 (a)	87 (a)
Myrrh	99	87 (a)
Myrtle	105 (a)	87 (a)
Nasturtium	105 (a)	87 (a)
Neat's-foot	101 (d), (e)	84
Neroli	105 (a)	87 (a)
Nutmeg	105 (a)	87 (a)
Olive	304	83
Orange	105 (a)	87 (a)
Orange berry	105 (a)	87 (a)
Orange flowers (Ol. Aurantii florum)	105 (a)	87 (a)
Orange leaf	105 (a)	87 (a)
Orange peel (Ol. Aurantii corticis)	105 (a)	87 (a)
Origanum (oil of thyme)	105 (a)	87 (a)
Orris root	105 (a)	87 (a)
Palm	100 (a)	83
Palma rosa	105 (a)	87 (a)
Partridge berry (see Wintergreen)	105 (a)	87 (a)
Patchouli	105 (a)	87 (a)
Peach kernels	105 (a)	87 (a)
Pennyroyal	99	87 (a)
Peppermint (Ol. Menthae piperitae)	105 (a)	87 (a)
Petit grain citronnier	105 (a)	87 (a)
Phosphorus (Ol. Phosphoratum)	99	81
Pimenta (see Allspice)	99	87 (a)
Pine (Ol. Pini)	105 (a)	87 (a)
Recini	100 (b)	83
Red cedar	105 (a)	87 (a)
Rhodium	105 (a)	87 (a)
Rose (Ol. Rosae; attar of roses)	105 (a)	87 (a)
Rose geranium	105 (a)	87 (a)
Rosemary (Ol. Rosmarini)	105 (a)	87 (a)
Rue	105 (a)	87 (a)
Sanitas	97	77
Santal, or sandalwood	99	87 (a)
Sassafras	105 (a)	87 (a)
Savine	99	87 (a)
Sesame	100 (b)	83
Sesamum	100 (b)	83
Shikima	105 (a)	87 (a)
Solvent		22
Spearmint	105 (a)	87 (a)
Sperm	101 (d), (e)	84

	1905. Par.	1909. Par.
Oils—Continued.		
Spike	99	87 (a)
Spruce	99	87 (a)
Star aniseed	105 (a)	87 (a)
Succini	99	87 (a)
Sweet birch	105 (a)	87 (a)
Sweet flag	105 (a)	87 (a)
Sweet hay	105 (a)	87 (a)
Tansy	99	87 (a)
Tar (Ol. Picis liquidae)	99	87 (a)
Tar, unrefined	7	22
Teaberry	105 (a)	87 (a)
Teeloil (see Sesamum oil)	100 (b)	83
Terebinthinae	77 (b)	63
Theobroma	290 (b)	248 (b)
Thyme	105 (a)	87 (a)
Tiglii	99.	83
Track	9	22
Tungtree	100 (b)	83
Turpentine	77 (b)	63
Valerian	99	87 (a)
Vegetable, fixed	100	83
Verbena	105 (a)	87 (a)
Vetiver essential	105 (a)	87 (a)
Violet	105 (a)	87 (a)
Volatile (see Essential)		87 (a)
Volatile sinapis	99	87 (a)
Walnut	100 (b)	83
Wax (Ol. Cerae)	99	84
Whisky	105 (a)	87 (a)
White cedar	105 (a)	87 (a)
White mustard	99	87 (a)
Wine	105 (a)	87 (a)
Wintergreen	105 (a)	87 (a)
Wood	99	83
Wormwood (Ol. Absinthi)	99	87 (a)
Oilcloth:		
In the piece	349	287 (a)
Made up into articles	349 (d)	287 (b)
Oil-extracting machinery	245	194
Oiled silk	174	147
Oilstones, whetstones, and hones of all kinds	1	3
Ointments	99	81
Oleaginous seeds	76	55 (a)
Oleic acid	92 (b)	71 (b)
Olein	101 (c), (e)	84
Oleographs	183	151
Oleomargarine	336	275
Olive oil	304	83
Olives:		
Fresh	374	234
Pickled—		
In bulk		236 (a)
In retail packages		236 (b)
Amnius gatherum		306
Onions:		
Canned	320	232
Fresh	375	317
Pickled	321	233
Onion-skin paper	179	150, 151
Onyx:		
In the rough	1 (a)	1 (a)
In slabs, plates, or steps	1 (b)	1 (b)
Beads, on strings in ornaments	340	279
Sculptures	1 (c)	1 (c)
Other articles	1 (d)	1 (c)

	1905. Par.	1909. Par.
Opals (*see* Precious stones)	24	13
Openers, can	46	194
Opera glasses. (According to material.)		
Opium, in whatever form		79
Prohibited, except as provided		Sec. 6 (g)
Opium pipes, prohibited		Sec. 6 (h)
Optical appliances and apparatus, not otherwise provided for		303
Orange:		
Berry, oil of	105 (a)	87 (a)
Essence, artificial	105 (a)	87 (a)
Flowers, oil of	105 (a)	87 (a)
Fruit—		
Conserved or crystallized		238
Dried	286	235
Fresh	374	234
Preserved	322	236
Leaf, oil of	105 (a)	87 (a)
Peel (dried)	81	57
Oil of	105 (a)	87 (a)
Water (not perfumery)	99	81
Orchilla extract	87 (a)	66 (b)
Ores	11, 369	309
Scoriae, from smelting of	75	309
Orexina, tannate	99	81
Organic:		
Acids	92	71
Salts	95	75
Salts, acid or double	97	77
Organs:		
Cabinet	236	185
Household effects of settlers coming to the Philippine Islands and returning residents thereof, free	393	341, 347
Pipe, for churches and societies		331
Toy mouth	353	296
Other		185
Origanum oil	105 (a)	87 (a)
Ornaments of common materials	340	279
Orphol	99	81
Orris-root oil	105 (a)	87 (a)
Orthoform	99	81
Osiers:		
Manufactured	204	169 (b), (c)
Unmanufactured	203	169 (a)
Ostrich feathers	230	290
Ostriches	212	175
Ownership of imported merchandise		Sec. 17
Oxalate of:		
Iron	95 (a)	75
Potassium	95 (a)	75
Oxalates	95 (a)	75
Oxalic acid	92 (a)	71 (b)
Oxen	208 (a)	171 (a), 333, 334, 342, 348
Ox gall, powdered	99	81
Oxide of:		
Antimony	99	81
Caustic and barilla alkalies	93	72
Cobalt	97	77
Copper	94 (e)	74 (c)
Iron	84 (c)	60 (c)
Lead	84 (a)	61 (a)
Manganese	83	60
Mercury	97	77
Nickel	97	77

	1905. Par.	1909. Par.
Pamphlets:		
For advertising, when free	381	327
Obscene or indecent, prohibited	Sec. 6	Sec. 6
Other	180	151, 154
Pancreatin	99	81
Panoramas for public entertainment, imported temporarily (bond taken for reexportation)	392	342
Pantanberge	99	81
Papain (payotin)	99	81
Papeles azoados	99	81
Paper:		
Albumen	187 (a)	151
Bags	178	150 (b), 151 (b)
Baryta coated	187	151
Basic, photographic, for albumenizing	187	150, 151
Bibulous	187	150, 151
Blotting	186	150, 151
Bond	179	150, 151
Books—		
Blank, printed or not	181, 181 (a)	153
Letterpress copying	187	153
Printed	180	154
Boxes	190	150 (b), 151 (b)
Canes		296
Carbon	187 (c)	151
Carborundum		149
Cardboard	189	150, 151
Cigarette	188	152
Coated with tar for roofing	348	149
Confetti and serpentines	190 (d)	150 (b)
Copying	187	150
Cover	187	150, 151
Drawing	187	150, 151
Embossed		151
Emery	185	149
Engraved		151
Envelopes	179	150 (b)
Etched		151
Filter	187	150
Fly, sticky or chemical	97	77
Gelatin coated	187	151
Gilt	187	151
Glass	185	149
Gum	187	151
Hats		296
Lace	190 (d)	151
Lanterns	190 (d)	151 (b)
Leaves for artificial flowers	350	292
Ledger	179	150, 151
Letter	179	150
Letterpress copying books	187	153
Lithographed		151
Lucky	187	150, 151
Manila	178	150, 151
Manufactures of	178, 198	Class IX
Napkins	190 (d)	150 (b), 151 (b)
Note	179	150
Old newspapers used for wrapping	178	151
Onion skin	179	150, 151
Otherwise elaborated		151
Parchment	187	150, 151
Patterns	190 (d)	150 (b), 151 (b)
Photographic—		
Plain basic	187 (b)	151
Sensitized	187 (b)	151
Pottery	187	150, 151

	1905. Par.	1909. Par.
Paving:		
Blocks of asphalt	7	22
Of stone (rough)	2 (a)	2 (a)
Slabs, cones, and prisms of glass for	14 (a)	18 (a)
Of stone for	2 (c)	2 (b)
Crushed stone for	2 (b)	2 (b)
Payment of duties	.	Sec. 9
Peaches:		
Dried	286	235
Fresh	374	234
Preserved	322 (b)	236
In cordials or spirits	323	238
Peach-kernel oil	105 (a)	87 (a)
Pea flour	287 (c)	229 (c)
Peanut oil	100 (b)	83
Peanuts	329	239
Conserved or crystallized	331	238
Pearl barley	282	221
Pearl, mother of	341	280
In buttons	345 (c)	283 (a)
Unwrought	341 (a)	280 (a)
Wrought	341 (b)	280 (b)
Pearl oyster shells, crude	366	284
Pearls and seed pearls	25	13
Imitations of	26	14
Pears:		
Fresh	374	234
Preserved	322 (b)	236
In cordials or spirits	323	238
Peas:		
Canned	320	232
Dried—		
In bulk	287 (a)	229 (a)
In small packages	287 (b)	229 (b)
Fresh	375	230
Pedometers and parts thereof		187
Pelletierine	96	78
Pelts (fur skins)	213	177
Pencils:		
Camel's hair	164	139
Copper sulphate in, for cauterizing	99	81
Lead, colored, charcoal, and indelible	85 (c)	65
Slate	2 (e)	2 (c)
Silver nitrate in, for cauterizing	99	81
Pen drawings	182	155, 326
Penholders. (According to material.)		
Penny royal oil	99	87 (a)
Pens:		
Fountain and parts and points for		295
Of common metals	51, 68 (c)	278
Other. (According to material.)		
Reservoir and parts and points for		295
Pepper:		
Ground	299 (b)	252 (b)
Whole	299 (a)	252 (a)
Peppermint oil	105 (a)	87 (a)
Pepper, oil, black	99	83
Volatile		87 (a)
Peppers:		
Fresh	375	230
Pickled	321 (a), (b)	233
Preserved		232
Peppers, pod, dried:		
Ground		252 (b)
Whole		252 (a)
Pepsin	99	81
Pepsine, elixir	99	81

	1905. Par.	1909. Par.
Peptone	99	81
Perambulators	261	201
Perchloride of iron, solid or solution of	97	77
Perfumery	105	87
Periodicals, public	382 (b)	328
Permanganate of potassium	97	77
Perouin	96	78
Peroxide of:		
Hydrogen	97	77
Manganese (so-called English iron black)	83	60
Sodium	97	77
Personal effects, when free	386, 393	339, 341, 347
Of settlers		341
Of tourists		339, 340
Of travelers		339, 340
Persulphate of sodium	94 (h)	74 (c)
Peru, balsam	99	81
Petit grain citronnier oil	105 (a)	87 (a)
Petrolatum (vaseline)	10	23 (c)
Petroleum:		
Crude	8	22
Rules for classification, of	Class II, Group 2	
Refined	9	23
Engines	243	194
Ether, chemically pure	97	77
Motors	243	194
Pewter. (*See* Zinc and alloys.)		
Pharmaceutical products not otherwise provided for	99	81
Defined		Sec. 8
Pharmacœpia or national formulary		Sec. 8
Pharmacy, substances employed in		Class IV
Philippine products, reimported		350
Philosophical and scientific apparatus, when free	390	349
Phonodoscope	365	303
Phonographs and graphophones, and detached parts for	254	186
Phosphate of:		
Iron (scales)	99	81
Lime	94 (d)	74 (a)
Sodium, granulated	97	77
Phosphide of zinc	99	81
Phosphoratum:		
Oil	99	81
Zinc	99	81
Phosphoric acid	91 (c)	70 (c)
Phosphorus	90 (b)	69
Phosphorous oil	99	81
Photo-engravings	182	151, 155, 326
Photographic:		
Albums, of paper	190 (d)	151, 155
Apparatus and equipment		302
Cameras	358, 359, 360	302
Card mounts	190	150, 151
Chemicals—		
Bromide of potassium	97	77
Chloride of—		
Gold	96	78
Silver	96	78
Euquinone	97	77
Hydroquinone	97	77
Hyposulphate of soda	94 (e)	74 (c)
Iodide of potassium	97	77
Metol	97	77
Nitrate of silver	94 (h)	74 (c)
Potassium chlorplatinite	97	77
Protargol	97	77
Sulphite of soda	94 (h)	74 (c)

	1905. Par.	1909. Par.
Photographic—Continued.		
Dry plates.	362	302
Films.	362	302
Flashlight powders, magnesium.	73 (f)	54 (b)
Lenses.	361, 362	302
Negatives.	362	302
Papers.	187	150, 151
Pictures.	182	151, 155, 326
Plate holders.	362	302
Tripods.	362	302
Photographs of actual persons.	380	302
Other.	182	151, 155
Pianolas.		185
Pianos and stools for.	235	185
Household effects, when free.	393	341, 347
Pickaxes.	46	194
Picks.	46	194
Picrotoxin.	99	81
Picture-projecting devices, parts, appurtenances, and accessories for.		186
Pictures:		
When free.	393	326, 337, 344
Obscene or indecent (prohibited).	Sec. 6	Sec. 6
Of actual persons.	380	326
Works of art.	380	326, 337, 344
Other.	182	155
Pigeons.	212, 268	175, 204
Clay, for targets.	19	11 (c)
Pig iron.	30	309
Pigments.		Class IV
Pigs, sucking.	210	173
Pile fabrics. (According to material.)		
Admixture of materials in.		Rule 6, Sec. 5
Pillow shams. (According to material.)		
Pills:		
Chinese.	98	80
Medicinal (except quinine).	98	80, 81
Quinine.	384	322
Pilocarpine.	96	78
Pimental oil.	99	81
Pini oil.	105 (a)	87 (a)
Pins:		
Of common metals.	52, 68 (c)	278
Other. (According to material.)		
Pipe organs for churches and societies.	195, 196, 197	331
Other.		185
Piperin.	99	81
Pipes:		
Aluminum.	71 (b)	52 (a)
Bowls for opium (prohibited).	23	Sec. 6
Cast-iron.	31 (a)	28 (a)
Clay.	17 (c), (d)	9
Copper.	67	48
Fittings for—		
Of cast iron.	31 (c), 32, 33	28 (b), 29
Of copper.	69 (a), (b), (c)	50
Of wrought iron.	58, 59	47
Lead.	73 (b)	54 (a)
Nickel.	71 (b)	52 (a)
Stove.	36 (c)	31 (b)
Tin.	72 (b)	53 (a)
Wooden receptacles for liquors.	363	161 (a)
Wrought iron and steel.	39 (a), (b)	35
Zinc.	73 (b)	54 (a)
Pipes, opium, prohibited.		Sec. 6 (h)
Piques.	122, 166	103
Pistils, artificial flowers.	350	292
Pistols, and parts for.	55 (c)	45

	1905. Par.	1909. Par.
Pitch:		
Burgundy	77 (a)	56 (a)
Felt, covered with, for roofing	348	286
Mineral	7	22
Paper, covered with, for roofing	348	149
Vegetable	77 (a)	56 (a)
Pitchers, glass	13	16
Pith, vegetable	366	169 (a)
Planes	46	194
Planing machinery	257 (b)	194
Planters, corn	245	194
Plants, live	367	308
Plaster of Paris	3 (a)	6 (b), (c)
Plasters, medicated	99	81
Plate glass	14	18 (c)
Plate holders and frames for cameras	362	302
Plates:		
Cast iron	31 (a)	28 (a)
Photographic, dry	362.	302
Steel, fish	38 (a)	33 (a)
Wrought iron	42	31 (a)
Platinum:		
In broken-up jewelry or table services, bars, sheets, dust, and scraps	372	318
Chloride	96	78
Jewelry	27	25 (a)
Other manufactures of	27	25 (d)
Playing cards, marked (prohibited)	Sec. 6	Sec. 6
Other	353	296
Pliers	46	194
Plows	245	194
Plumbago	89	67
Plumbers' knives	54 (a)	44 (a)
Plum pudding	325	224
Plushes. (According to material.)		
Admixtures of materials in		Rule 6, Sec. 5
Pneumatic:		
Pumps	257	194
Tires for—		
Automobiles	352 (d)	198
Bicycles	252	199
Pocketbooks. (According to material.)		
Pocket cutlery	54 (b)	44 (b)
Podophyllom	99	81
Pod peppers, dried:		
Ground		252 (b)
Whole		252 (a)
Poker chips. (According to material.)		
Police, messenger, and fire-alarm apparatus, electric	248	193 (a)
Polish, shoe	89	67
Poligalae	99	81
Polo mallets		297
Pomades	105 (b)	87 (b)
Pongee silk	174	147
Pool:		
Balls		166
Tables, parts, and appurtenances for		166
Poppy heads, dried	81	57
Porcelain wares	21, 22, 23	11
Buttons	345 (a)	283 (b)
Insulators	248	193 (a)
Pork:		
Canned or potted	315, 316	209
Fresh	376	205
Salted or in brine	269	206
Smoked or cured	270	207

	1905. Par.	1909. Par.
Porous plaster	99	81
Portable engines	245, 246	194
Portieres. (According to material.)		
Portland cement	3 (a)	6 (a)
Postage stamps	183 (a)	329
Postal cards, pictorial	179	151 (b), 155
Posters, advertising, when free	381	327
Potash (caustic alkalies)	93	75
Salts of. (*See* Potassium.)		
Potassium:		
Acetate	95 (b)	72
Bicarbonate of	94 (h)	74 (c)
Bichromate of	94 (h)	74 (c)
Bitartrate (cream of tartar)	97	76
Bromide of	97	77
Carbonate of	94 (h)	74 (c)
Chlorate of	94 (f)	74 (c)
Chloride of	94 (c)	74 (a)
Chlorplatinite	97	77
Citrate of	95 (b)	75
Cyanide of	97	77
Ferrocyanide	97	77
Glycerophosphate of	99	81
Hydroxide of	93	72
Hypophosphate of	99	81
Iodide of	97	77
Nitrate of	94 (d)	74 (a)
Nitrite of	94 (h)	74 (c)
Oxalate of	95 (a)	75
Oxides of	93	72
Permanganate of	97	77
Prussiate	97	77
Sulphate of	94 (e)	74 (a)
Sulphite of	94 (h)	74 (c)
Sulphuret	94 (h)	74 (c)
Tartrate	95 (b)	75
Potatoes, Irish, free		317
Pots, flower:		
Of common clay and earthenware, plain	19 (c)	11 (b)
Ornamental, of faience, fine clay, porcelain, stoneware, or bisque	23	11 (f)
Other. (According to material.)		
Pottery (*see* Clay), faience, porcelain	19, 20, 21, 22, 23	11
Pottery paper	187	150, 151
Poultices		81
Poultry, live	212	175
Other	268	204
Powder:		
Bronze	69 (a)	50
Egg		272
Emery	3 (b)	6 (b), (c)
Flashlight	73 (f)	54 (b)
Gun and miners' (explosives)	111	91
Prohibited		Sec. 6 (a)
Insect, vegetable (dalmatian)	81	57
Insect, chemical	97	77
Milk		271
Puffs—		
Of feathers	231 (a)	291 (b)
Of wool	166	143
Sachet	105 (b)	87 (b)
Sienna in	83	60
Talcum, natural	3 (b)	6 (b), (c)
For toilet purposes	105 (b)	87 (b)
Thoriae in	99	81
For toilet use	105 (b)	87 (b)
Tooth	105 (b)	87 (b)

	1905. Par.	1909. Par.
Powders:		
Baking	97	77
Explosive—		
For mining	111 (a)	91 (a)
Other	111 (b)	91 (b)
Seidlitz	99	81
Soap	104	86
Pozzuolano earth	3 (b)	6 (b), (c)
Precious stones:		
Cut, or uncut, unset	24, 379	13
Imitations of	26	14
Precipitated chalk	3 (a)	6 (c)
Preservers, life, free		310
Presses, printing, all kinds	257 (b)	194
Primers for firearms	346	91 (b)
Printed:		
Blank books	181 (a)	153
Matter for advertising, when free	381	327
Music, or lithographed sheet	180 (a)	151
Other printed matter	180	154
Printing:		
Ink	85 (a)	64 (a)
Paper	177	149
Presses	257 (b)	194
Prisms, of glass, for paving		18 (a)
Prizes, when free		330
Products, domestic, imported	Sec. 18	350
Products, edible, not otherwise provided for:		
Crude		276 (a)
Other		276 (b)
Prohibited importations	Sec. 6	Sec. 6
Projecting devices, parts and appurtenances therefor, for enlarging pictures	254	186
Proof liter for tariff purposes, defined		Class XIII, Group 6
Propellers for vessels	257	310
Propylamine	99	81
Proprietary defined		Sec 8.
Proprietary medicines	99	80
Protalgol	97	77
Pruning knives and shears	54 (c)	44 (a)
Prussiate of potassium	97	77
Publications, magazines, and periodicals:		
Free	382	328
Offensive to morality	Sec. 6 (2)	Sec. 6
Other		151, 154
		329
Public documents, issued by foreign governments, free	325	224
Puddings	245	194
Pulleys:		
Differential	245	194
Fine wood	196	163
Ordinary wood	195	162
Steel, plain	58	47
Pulp:		
Fruit	313	237
Paper	176	315
Pulp board, strawboard, pasteboard, and cardboard		150, 151
Manufactures of	150 (b), 151 (b)	156
Pulp or fiber, indurated		
Pulse:		
Dried—		
In bulk	287 (a)	229 (a)
In retail packages	287 (b)	229 (b)
Flour	287 (c)	229 (c)
Pulsometers (a kind of pump)	257 (b)	194
Pumice	3 (b)	6 (b), (c)
Pumpkin fiber (loofah sponges)	365	285
Pumpkins. (*See* Vegetables.)		

	1905. Par.	1909. Par.
Pumps	257	194
Intended for salvage of vessels, imported temporarily	396	346
Punk paper	187	150
Pure Food Law, articles violating the provisions of, prohibited		Sec. 6 (e)
Push buttons, electric	248	193 (a)
Putty	84 (b)	61 (b)
Putz pomade	97	77
Pyramidon	99	81
Pyridin	99	81
Quadrants	365	303
Quarterolas	363	161 (a)
Queensware (see Stoneware)		11
Quill toothpicks	342 (b)	281 (b)
Quina ferruginated	99	81
Quinidine sulphate	384	322
Quinine bark, sulphate and bisulphate of	384	322
Rabbits' feet, mounted in brushes	365	290
Rackets, tennis		297
Radiometers		303
Rahi (Indian spice)	296	255 (b)
Rails, wrought iron	35 (a)	30 (a), (b)
Bent or bolted in crossings, etc. (see note)	35 (a)	30 (a), (b)
Railway:		
Axles, wrought iron or steel	38 (a)	33 (a)
Carriers, public or common	262	200 (a)
Other		200 (b)
Chairs, wrought iron or steel	38 (a)	30 (a),)b)
Fish plates, wrought iron or steel	38 (a)	30 (a), (b)
Frog switches (see note)	42	30 (a)
Locomotives	246	194
Lubricating boxes, cast iron	31 (b)	28 (b)
Wrought iron or steel	38 (a)	33 (a)
Rails, wrought iron or steel	35 (a)	30 (a)
Sleepers, wrought iron or steel	38 (a)	30 (a), (b)
Springs	38 (a)	33 (a)
Switches (see note)	42	30 (a)
Tenders, for locomotives	246	194
Turntables	247	194
Wheels, wrought iron or steel—		
Weighing more than 100 kilos	38 (a)	33 (a)
Weighing 100 kilos or less	38 (b)	33 (b)
Raisins, dried:		
In bulk	286	235 (a)
In retail packages	285	235 (b)
Rakes	46	194
Ramie		Class VI
Raw or hackled	138	313
Rope or cordage	146	120
Textiles		Class VI, Group 2
Thread, twine, cord, and yarn	144	119, 120
Rasps	46	194
Rat poison	97	77
Ratchet braces	46	194
Rates of duty:		
Establishment of		Sec. 11
Two or more applicable, how assessed		Rule 12, Sec. 5
Rattan		169
Rattles, baby (toys)	353	278
Rawhide, manufactures of		181 (b)
Razor:		
Hones	1 (d)	3
Strops of leather	229	182
Razors	54 (b)	44 (b)
Safety	54 (b)	44 (b)
Reading glasses	16 (c)	20 (e)

	1905. Par.	1909. Par.
Rhoestats...	250	193 (a)
Rhodium oil...	105 (a)	87 (a)
Rhubarb root..	81	57
Ribbons:		
Admixtures of materials in........................		Rule 7, Sec. 5
Of cotton.......................................	131	113
For cinches or saddle girths....................	134 (a)	116
Of linen, hemp, flax............................	157	131
For cinches or saddle girths....................	159 (a)	134
Of mixed materials..............................	Rule 6	Rule 7
Of silk...	174	148
For typewriters.................................	253	188
Of wool...	166	143
Rice:		
Cooked...	282	221
Hullers..	245	194
Flour..	275 (c)	218 (c)
Husked...	276 (b)	218 (b)
Unhusked.......................................	276 (a)	218 (a)
Machinery for preparing.........................	245	194
Papers for cigarettes...........................	188	152
Powder for toilet use...........................	105 (b)	87 (b)
Red..	87 (b)	276 (b)
Threshers......................................	245	194
Ridgings, iron.......................................	36	31 (b)
Riding boots..	226	180
Rifles and parts for.................................	55, 56	45
Rings, button.......................................		278
Rivets:		
Copper...	68 (a), (b)	50
Iron and steel..................................	47	39
Road-making machinery..............................	245	194
Rochelle salts.......................................	97	77
Rock crushers.......................................	245	194
Rock sugar..	289 (b)	240 (b), (c)
Rodinal...	97	77
Roe, fish, canned....................................	318	215 (c)
In bulk..		216 (b)
Roller composition (composed of glue and gelatine, for		
printing press).................................	365	90
Roneos, parts of and accessories for..................		188
Roofing:		
Felt, tarred or covered with pitch, for...........		286
Glass slabs, cones and prisms, for...............	14 (a)	18 (a)
Paints, bituminous..............................	85 (a)	61 (b)
Paper, tarred or covered with pitch, for..........	348	149
Rubberoid and similar materials, for.............	348	286
Root beer...	312 (c)	268
Extract for making..............................	327	258
Root of:		
Angelica, alcanna, galangae, gentian, ipecacunha,		
iris, licorice, rhubarb, squilla, etc.............	81	57
Ginseng..	81 (a)	58
Roots for dyeing or tanning..........................	86	66 (a)
Rope:		
Cotton...	133	96
Flax, hemp, jute, and other vegetable fibers.......	146	120
Ropemakers' wares or vegetable fibers other than cotton..	146	120
Rose flowers, dried..................................	81	57
Rose geranium oil...................................	105 (a)	87 (a)
Rosmarina oil.......................................	105 (a)	87 (a)
Rosemary:		
Leaves...	81	57
Oil..	105 (a)	87 (a)
Rose oil..	105 (a)	87 (a)
Rosettes, electrical.................................	248	19?

	1905. Par.	1909. Par.
Rotary pumps	257	194
Rotten stone	3 (b)	6 (b), (c)
Rouge, toilet preparation	105 (b)	87 (b)
Jewelers'	97	77
Roulette wheels (prohibited)	Sec. 6	Sec. 6 (c)
Rubber:		
Balls, for children (toys)	253	296
Bands	352 (d)	293 (b)
Belting for machinery	352 (e)	196
Boots and shoes	352 (c)	180 (b)
Buttons	345 (b)	283 (c)
Combs	339 (a)	293 (c)
Gaskets	352 (a)	293 (a)
Hose	352 (e)	294
Hard rubber articles	352 (b)	293 (c)
Matting, corrugated	352 (d)	293 (b)
Packing for machinery	352 (a)	293 (a)
Raw or in lumps	77 (c)	293 (a)
Sponges		293 (b)
Tires for—		
Automobiles	352 (d)	198
Bicycles	252	199
Other vehicles		293 (b)
Toys	353	297
Washers	352 (a)	293 (a)
Other manufactures of	352 (d)	293 (b), (c)
"Rubberoid," and similar materials for roofing and structural purposes	348	286
Rubian	86 (b)	66
Rue oil	105 (a)	87 (a)
Rugs. (According to material.)		
Rules and regulations		Sec. 27
Rules, general		Sec. 5
Rum	308 (a)	263
Rum and quinine hair tonic	105 (b)	87 (b)
Rushes:		
Manufactured	204	169 (c)
Unmanufactured	203	169 (a)
Russet shoe polish	89	67
Russian caviare	318	215 (c)
Rye:		
In flour	277 (b)	219 (b)
In grain	277 (a)	219 (a)
Whisky	308 (a)	263
Saccharine	326	243
Saccharum lactis	99	81
Sachet powder	105 (b)	87 (b)
Sacks:		
Common (except gunny sacks), making up, surtax for		Rule 11, sec. 5
Containers of imported merchandise	391	355
Gunny	145	121
Other. (According to material.)		
Saddle:		
Blankets of wool	166	143
Girths—		
Of cotton	134	115
Of linen, hemp, and other vegetable fibers	159	133
Of wool	166	143
Saddlery	228	181 (b)
Hardware (wrought iron or steel)	49	27 (b), 42
Saddlers' knives	54 (a)	44 (a)
Saddles	228 (c)	181 (b)
Safes:		
Of cast iron	32, 33	29
Of wrought iron or steel	58, 59	47
Of wood	195, 196, 197	162, 163, 164

	1905. Par.	1909. Par.
Safety:		
Pins—		
Of common metals	52, 68 (c)	278
Other. (According to material.)		
Razors	54 (b)	44 (b)
Saffron	296 (c)	254
Sage:		
Ground	286 (b)	255 (b)
Leaves	81	57
Sago	284	225
Sake, Japanese wine		263, 264, 265, 266
Salacetol	99	81
Salad oil	304, 305	83
Salamander (asbestos)	3 (b)	4
Saleratus	97	77
Salicin	99	81
Salicylate:		
Of cerium	99	81
Of bismuth	99	81
Of soda, sirup of	99	81
Saline effervescent	99	81
Sallis water, hair dye	97	87 (b)
Salmometers		303
Salmon:		
Fresh	274 (a)	216 (a)
Salted, smoked or pickled, in bulk	274 (b)	216 (b)
Canned or potted	317 (a)	215 (a)
Salofeno	99	81
Salol	99	81
Salophen	99	81
Sal soda	94 (e)	74 (c)
Salt:		
Carlsbad	99	81
Common, ground, "sodium chloride"	94 (b)	74 (c)
Fruit	99	81
Glaubers'	94 (c)	74 (c)
Sorrel	95 (a)	75
Used for packing hams	94 (a), (b)	74 (c)
Water soap	104	86
Wiesbaden	99	81
Salted:		
Eggs, in natural form	333	272 (a)
Fish—		
Codfish	273	214
Sea blubber	274 (b)	216 (b)
Stockfish	273	214
Other	274 (b)	216 (b)
Meats	269	206
Saltpeter	94 (d)	74 (c)
Salts:		
Of alkaloids, except opium	96	78
Of opium, alkaloids		79
Of cinchona	384	321
Inorganic	94	74
Alum	94 (c)	74 (c)
Bicarbonate of sodium	94 (g)	74 (c)
Borax	94 (e)	74 (c)
Carbonate of ammonium and magnesium	94 (e)	74 (c)
Chloride of ammonium	94 (e)	74 (c)
Calcium carbide	94 (e)	74 (c)
Calcium hypochlorite		74 (c)
Chlorate of potassium	94 (f)	74 (c)
Chlorate of sodium	94 (f)	74 (c)
Chloride of lime	94 (e)	74 (b)
Chloride of potassium	94 (c)	74 (a)
Chloride of sodium	94 (a), (b)	74 (c)
Epsom salt (sulphate of magnesium)	94 (c)	74 (c)

	1905. Par.	1909. Par.
Salts—Continued.		
Inorganic—Continued.		
Glauber salt	94 (c)	74 (c)
Hyposulphite of sodium	94 (e)	74 (c)
Nitrate of copper	94 (e)	74 (c)
Nitrate of potassium	94 (d)	74 (a)
Nitrate of sodium	94 (d)	74 (a)
Other, not specially provided for	94 (h)	74 (c)
Oxide of copper	94 (e)	77
Phosphates of lime	94 (d)	74 (a)
Sal soda	94 (e)	74 (c)
Sulphate of ammonium	94 (d)	74 (a)
Sulphate of copper	94 (e)	74 (c)
Sulphate of potassium	94 (e)	74 (a)
Sulphate of iron	94 (c)	74 (c)
Sulphate of magnesium	94 (c)	74 (c)
Sulphate of sodium	94 (c)	74 (c)
Organic—		
Acetates	95 (a)	75
Citrates	95 (b)	75
Oxalates	95 (a)	75
Tartrates	95 (b)	75
Rochelle	97	77
Smelling (perfumed)	105 (b)	87 (b)
Salvage of vessels, pumps for	396	346
From vessels built in foreign countries		Rule 12, sec. 5
Salve	99	81
Samples:		
Commercial, in bond, free		335
Of no appreciable value	370 (a)	317
Usual commercial	370	335
Sand	3 (b)	6 (b), (c)
Paper	185	149
Sandal-wood oil	99	87 (a)
Sandals	227	180
Sandalwood		57, 158
Chips	81	57
Oil	99	87 (a)
In capsules, for medical use		81
Sandarac, gum	78	56 (b)
Sanitary construction articles:		
Rates of duty on		Sec. 11
Of plain lead for		54 (a)
Sanitas oil, disinfectant	97	77
Sanmetto	99	81
Santal:		
Oil of	99	87 (a)
In capsules for medicinal use		81
Wood, red, for dyeing	86 (a)	66 (a)
Sapolio	104	86
Saponified resin	104	86
Sapo viridis	104	86
Sarongs, making up defined, surtax for		Rule 11, sec. 5
Sarsaparilla:		
Beverage	312 (c)	268
Drug	99	81
Root	81	57
Sassafras	81	57
Oil of	105 (a)	87 (a)
Sardines	317	215 (a), (c)
Satsuma ware	23	11 (f)
Buttons of	345 (a)	283 (b)
Sauces, for table use	328	256
Sauerkraut:		
In retail packages	320	232 (b)
In bulk	321 (b)	232 (a)
Saucers. (According to material.)		

	1905. Par.	1909. Par.
Sausage casings		206
Sausages	270, 315, 316	207, 210
Savon	104	86
Sawdust (common)	366	159
Sawmill machinery and detached parts for	245	194
Saws	46, 245, 257 (b)	194
Scales:		
Copper (laminæ)	60	309
For weighing, and detached parts for	242	192
Schists	7	Class II, Group 2
Crude oils derived from	8	22
Refined illuminating oils derived from	9	23 (a)
Refined oils, other, derived from	10	23 (b), (c), (d)
School:		
Books (text)	382 (b)	328
Slates	2 (e)	2 (c)
Scientific:		
Books, apparatus, utensils, and instruments specially imported in good faith, for use and by the order of any society established solely for scientific purposes	390 (a)	349
Apparatus and appliances not otherwise provided for		303
Scissors:		
Plain, glazed, or japanned	54 (a)	44 (a)
Pocket	54 (b)	44 (b)
Surgical	54 (e)	196
Other	54 (d)	44 (b)
Scopolamine	96	78
Scoriæ, resulting from smelting of ores	75	309
Scouring compositions	104	86
Scraps:		
Brass, copper, iron, lead, tin, and zinc		309
Gold, silver, and platinum	372	318
Of other metals	74	309
Screens. (According to material.)		
Screw drivers	46	194
Screws:		
Bench	46	194
Copper and brass	68 (a), (b)	50
Hand	46	194
Jack, hoisting apparatus	245	194
Wrought iron or steel	47	41
Sculptures, when free		336, 337, 344
Of marble, jasper, onyx, etc	1 (c)	1 (c)
Of artificial or common stone	2 (e)	2 (c)
Of clay, faience, porcelain	23	11 (f)
Of gypsum	4 (a)	7
Scythes	46	194
Sea blubber, salted	274 (b)	216 (b)
Seaweed, dried, edible	288	276
Crude		276 (a)
Advanced in condition		276 (b)
Sealing wax	97	77
Seal presses of cast iron	33	29
Sedlitz:		
Granules and powders	99	81
Mixture	99	81
Seed pearls	25	13
Seeds:		
Anise	81	57
Aromatic and also of morbid growth	81	57
Artificial	350	292
Canary	302	227
Celery	296 (a)	255

	1905. Par.	1909. Par.
Seeds—Continued.		
Coriander and caraway	81	57
Cotton	76	55 (a)
Ground, for culinary purposes (spices)	296 (b)	255 (b)
Flaxseed	76	55 (a)
Flower	302	227
Garden and grass	302	227
Linseed	75	55 (a)
Medicinal	81	57
Mustard	300	253 (a)
Not otherwise pro ided for	302	227
Oleaginous v	76	55 (a)
Paeonia	81	57
Sesame	76	55 (a)
Strophantus	81	57
Seines. (*See* Nets.)		
Semi-precious stones	26	13
Senar blacking	89	67
Senegal gum	78	56 (b)
Senna leaves	81	57
Sensen:		
Breath perfumery	105 (b)	87 (b)
Chewing gum	332	244
Sensitized paper	187 (b)	151
Separators, grain	245	194
Serpentine paper and confetti	187	150 (b)
Serums and vaccines		324
Sesame seed	76	55 (a)
Oil of	100 (b)	83
Sets, croquet		297
Settlers, personal and household effects of, when free		341
Sevres wares, not otherwise provided for		11 (f)
Sewing:		
Machine needles	51	278
Machines and detached parts (except needles)	251	190
Needles	51	278
Thread—		
Cotton	116	95
Hemp, jute, and other vegetable fibers	144	120
Silk	170, 172	146
Sextants	365	303
Shafts, wooden, for vehicles	266	202
Sharks' fins	274 (b)	215 (c)
Shaving soap	104	86
Shavings of iron and common metals	74	309
Shawl straps of whatever material	229	288
Shawls. (According to material.)		
Shawls (mantones and panolones), making-up defined, surtax for		Rule 11, Sec. 5
Shears:		
Garden, grass, hedge, and pruning	54 (c)	44 (a)
Plain, glazed, or japanned	54 (a)	44 (a)
Sheep	54 (c)	44 (a)
Other	54 (d)	44 (b), 194
Sheath knives		44 (b)
Sheep	211	174
Shears	54 (c)	44 (a)
Skins—		
(Basils) curried	217 (a)	178 (a)
Known to the trade as chamois skin	219	178 (c)
Skin shoes	221	180 (a)
Shell, tortoise, and mother-of-pearl:		
Unwrought	341	280 (a)
Wrought	339 (b)	280 (b)
Other, export duties on		363
Shellac	78	56 (b)

	1905.	1909.
	Par.	Par.
Shellfish:		
Canned or potted	317, 318	215 (b)
Fresh oysters, in cans	275	217
Other	275	217
Shells, oyster, crude	366	284 (a)
Crude, polished or not		284 (a)
Export duties on		363
Further advanced in condition		284 (b)
Further advanced in condition		285 (b)
Shikimi oil	105 (a)	87 (a)
Shingles, laths, and fencing, common wood	192 (a)	157 (b)
Ship augers	46	194
Shipping tags	190 (a), (d)	150, 151
Ships' logs (instruments)	257 (a)	303
Shirts. (According to material.)		
Shoe:		
Dressing	89	67
Eyelets of brass for	69 (b)	50
Findings		178
Laces. (According to material.)		
Polish	89	67
Shoemakers':		
Heelball	103	85 (c)
Knives	54 (a)	44 (a)
Shoes:		
Blanco for	89	67
Eyelets, or eyelet rings for, of brass	69 (a), (b)	50
Of calfskin	223	180 (b)
Of canvas	221	180 (a)
Of cowhide	221	180 (a)
Of goatskin, and lambskin	223	180 (b)
Of kidskin	223	180 (b)
Of patent leather or imitation patent leather	222	180 (b)
Of rubber	352 (c)	180 (b)
Of sheepskin	221	180 (a)
Of varnished canvas, imitating patent leather	222	180 (b)
Other, not otherwise provided for	225 (a)	180 (b)
Polish or dressings for	89	67
Riding boots	226	180
Silk		180 (c)
Whiting for (blanco)	89	67
Wooden	227 (b)	180 (b)
Shooks, staves, headings, hoops, and bungs, wood	194 (b)	160
Shoots, plants, trees, and moss, live	367	308
Shot and bullets of lead		54 (b)
Shotguns and parts for	56	45
Cartridges for	346	91 (b)
Shovels	46	194
Wooden handles for	194 (e)	194
Show cases. (According to material.)		
Shredded:		
Cocoanut	331, 365	239
Wheat	282	221
Side arms:		
(Not fire) and detached parts for	54 (b)	44 (b)
Pistols and revolvers, and detached parts for	55 (c)	45
Sienna, earth:		
Dry	83	60
In liquid or paste	84 (d)	61 (d
Sieves. (According to material.)		
Signals, explosive		91 (a)
Signs:		
Advertising, when free		327
Other. (According to material.)		
Sight-testing, appliances for		303

	1905. Par.	1909. Par.
Silk:		
Cocoons and eggs of silkworm	167, 168	332
Floss	171	146
Knitted	173	148
Ligatures for surgical use	99	82
Manufactures of	196–175	148, Class VIII
Oiled (in the piece)	174	147
Spun, not twisted	169	145
Strings for musical instruments	174	185
Textiles (in the piece)	174	147
Trimmings and lace	174	148
Thread, twisted	170	146
Waste		144
Worm eggs	167	332
Yarns	169–172	146
Silver:		
Baby rattles	28 (d)	26 (d)
Chloride of	96	78
Coins of national issues		320
Combs	28 (a), (d)	26 (a), (d)
Cups, as prizes or trophies	28 (a)	330
Foil and pellets for dentists	28 (d)	26 (d)
German (usually copper alloy). (See Copper.)		
Hair brushes	28 (d)	26 (d)
Ingots, broken-up jewelry, or table service, bars, pieces, dust, and scraps	372	318
Inkstands	28 (d)	26 (d)
Jewelry, plate and toilet articles, not otherwise provided for	28 (a)	26 (a)
Set with pearls or precious stones	28 (b)	26 (b)
Set with imitation pearls or precious stones	28 (c)	26 (c)
Leaf	28 (d)	26 (d)
Letter cases	28 (d)	26 (d)
Medals	28	26, 330
Mirrors	28 (d)	19
Nitrate, crystals	97	77
For cauterizing	99	81
Ore	369	309
Pellets and foil, for dentists		26 (d)
Plate	28 (a)	26 (a)
Plated—		
Jewelry	29 (a)	27 (a)
Lamps, picture frames, knives, forks, and spoons, carriage, and coffin fittings, saddlery hardware.		27 (b)
Other wares		27 (c)
Strings for musical instruments		185
Solder	28 (d)	26 (d)
Thread	28 (d)	26 (d)
Wire	28 (d)	26 (d)
Other articles	28 (d)	26 (d)
Sinapis oil, volatile	99	87 (a)
Singing birds	212	175
Siphons, ordinary glass, for aerated waters	12 (b)	15 (b)
Sirup:		
Flavoring		258
Maple		241
Medicinal	99	81
Of fruits	313	258
Sugar	337	241
Skiffs	267	203
Skins, hides, and leathers:		
Bronzed or gilt	219	178 (c)
Chamois leather, parchment, and vellum	219	178 (c)
Curried	217	178
Sheepskins (basils)	217 (a)	178 (a)
Calf, goat, kid or lamb	217 (b)	178 (b)
Cow	217 (d)	178 (a)

	1905. Par.	1909. Par.
Skins, hides, and leathers—Continued.		
Enameled, gilt, bronzed, bleached, figured, engraved or embossed (fur skins)		178 (c)
Green, raw or dry	214	176
Pelts (fur skins)	213	177
Tanned, with the hair on	215	177
Tanned without the hair	216	178
Slate:		
Pencils	2 (e)	2 (c)
In slabs, plates, or steps	2 (c)	2 (b)
Wrought into other articles	2 (e)	2 (c)
Unwrought	2 (a)	2 (a)
Slates, imitation (roofing)	2 (c)	2 (b)
Slates, school	2 (e)	2 (c)
Sledge hammers	46	194
Sleepers or ties:		
Of cast iron	31 (a)	28 (a)
Of ordinary wood	192 (a)	157 (a)
Of wrought iron or steel	38 (a)	30 (a),(b)
Slides, glass, for magic lanterns	254	186
Slippers:		
Of silk	225	180 (c)
Other	227	180 (a), (b)
Sloe gin	308 (c)	263
Slot machines:		
Automatic, for weighing and other purposes not prohibited, and detached parts for	256	186, 191
Gambling, prohibited	Sec. 6	Sec. 6
Small arms (fire), and detached parts for	55	45
Smoked:		
Fish	174 (b)	216 (b)
Meats	270	207
Smoothing irons of cast iron	32, 33	29
Electric		193 (b)
Snap switches, electric	248	193 (a)
Snatch blocks, steel, plain	58	47
Snuff (tobacco)	364 (b)	304 (b)
Medicinal		81
Soap	104	86
Bark	81	57
For dyeing (Maypole)	87 (b)	66 (c)
Powders	104	86
Salt water	104	86
Shaving	104	86
Soft	104	86
Tooth	104	87 (b)
Viridis	104	86
Sockets, electrical	248	193
Socks. (According to material.)		
Soda:		
Ash	94 (e)	72
Biscuit	283 (a)	223
Cooking	94 (g)	74 (c)
Sal soda	94 (e)	74 (c)
Sirup salicylate	99	81
Washing	94 (e)	74 (c)
Sodium:		
Acetate	95 (a)	75
Arseniate	99	81
Bicarbonate	94 (g)	74 (c)
Bisulphite	94 (h)	74 (c)
Bromide	97	77
Carbonate (sal soda)	94 (e)	74 (c)
Chlorate	94 (f)	74 (c)
Chloride (common salt)	94	74 (c)
Citrate	95 (b)	75
Glycerophosphate of	99	81
Hyposulphite	94 (e)	74 (c)

	1905. Par.	1909. Par.
Sodium—Continued.		
Nitrate	94 (d)	74 (a)
Oxide and hydroxides	93	72
Peroxide	97	77
Persulphate	94 (h)	74 (c)
Phosphate	94 (h)	74 (c)
Salicylate	99	81
Sulphate	94 (c)	74 (c)
Sulphite	94 (b)	74 (c)
Soft soap	104	86
Solanine	96	78
Solder:		
Gold, for dentists' use	27 (d)	25 (d)
Lead, in bars, etc	73 (b)	54 (a)
Silver	28 (d)	26 (d)
Tin, in bars	72 (a)	53 (a)
Zinc, etc., in bars	73 (b)	54 (a)
Solvent oils		23 (b)
Somatose	99	81
Soot	89	61
Sorrel salt (potassium oxalate)	95 (a)	75
Sounds, fish		183
Soup:		
Canned	319	212
Condensed or concentrated, preparations of		213
Pastes for	284	225
Soy	328	256
Spades	46	194
Wooden handles for	194 (e)	194
Sparkling wine	309	264
Sparteine	96	78
Spearmint oil	99	87 (a)
Specimens of mineralogy, botany, zoology, and ethnology, and small models for public museums, public schools, academies, and scientific and artistic societies	389	344
Spectacles of all kinds and glasses for the same	16 (a)	20 (a)
Speculums, surgical	54 (e)	303
Spelter (solder), (copper brazing compound)	69 (a)	50 (a)
Sperm oil	101 (d), (e)	84
Spermaceti	101 (c), (e)	84
Spices:		
Allspice and mace—		
Ground	297 (b)	250 (b)
Unground	297 (a)	250 (a)
Cinnamon and cloves—		
Ground	293 (b), 294 (b)	250 (b)
Unground	293 (a), 294 (a)	250 (a)
Indian Jiri, Hurbo, and Rahi	296	255
Nutmegs—		
Husked	298 (b)	251 (b)
Unhusked	298 (a)	251 (a)
Ground	296 (b)	251 (c)
Not otherwise provided for—		
Ground	296 (b)	255 (b)
Unground	296 (a)	255 (a)
Saffron	296 (c)	254
Spigle Yohimbina	96	78
Spigots, copper	69	50
Spike oil	99	87 (a)
Spirit levels	46	194
Spirits:		
Compound	308 (c)	263
Of turpentine	77 (b)	63
Splice bars, steel, fish plate	38 (a)	33 (a)
Sponges	365	285
Rubber	352 (d)	294 (b)
Hexactinellida		285
Loofahs (pumpkin or vegetable fiber)	365	285

	1905.	1909.
	Par.	Par.
Spools. (According to material.)		
Spoons. (According to material.)		
Springs:		
For carriages and wagons, of wrought iron or steel....	38 (b)	33 (b)
For furniture, of wrought iron or steel...............	45	47
For railways or tramways, or wrought iron or steel...	38 (a)	33 (a)
Spruce oil...	99	87 (a)
Spun silk...	169	145
Spuriously stamped or marked articles of gold or silver, or their alloys, prohibited.........................		Sec. 6 (d)
Spurs:		
Of copper..	69 (a), (b)	50
Of wrought iron or steel............................	49 (a), (b)	42
Others. (According to material.)		
Squares:		
Carpenters'..	46	194
Of clay...	17 (a), (b)	9
Spuilla root..	81	57
Stallions...		170, 333, 342, 348
Stamp:		
Mills..	257 (b)	194
Pads..	85	188
Stamped falsely, articles of gold or silver, prohibited....		Sec. 6 (d)
Stamps, postage or revenue, canceled or not, of national issues..	183	329
Stands:		
For clocks...	239	187
Ink. (According to material.)		
Flower, of ceramic materials........................	23	11 (f)
Others. (According to material.)		
Staples:		
For fastening paper. (According to material.)		
Star aniseed oil..	105 (a)	87 (a)
Starch:		
Corn, for table use.................................	282	221
For industrial purposes.............................	107	89
Stationary engines.....................................	243	194
Statuetts:		
Of bisque, clay, faience, porcelain, stoneware, etc...	23	11 (f)
Of crystal and glass imitating crystal................	13 (a), (b)	20 (b), (c)
Of gypsum..	4 (a)	7
Other. (According to material.)		
Staves, headings, hoops, bungs, and shooks.............	194 (b)	160 (b)
Stearic acid..	92 (b)	71 (b)
Stearin:		
In candles..		85 (b)
Unwrought...	101 (c)	85 (a)
Wrought in other articles...........................	103	85 (c)
Steel and wrought iron:		
Bolts..	47	39
Butchers' tools......................................	46	194
Buttons...	345 (a)	283 (b)
Crucible, in bars, beams, and rods..................	35 (c)	30 (d
Cutlery..	54	44
Frogs, railway switch...............................	42	30 (a), (b)
Ingots...	34	309
Mooring buoys......................................	42	47
Nuts...	47	39
Rivets...	47	39
Screws..	47	41
Sheets, rolled.......................................	36 (a)–(c)	31 (a), (b)
Springs—		
For wagons and carriages.........................	38 (b)	33 (b)
For railways and tramways........................	38 (a)	33 (a)
Other...	45	47
Structural..	41	34
Tapes for measuring.................................	46	194

	1905. Par.	1909. Par.
Steel and wrought iron—Continued.		
Tools and implements	46	194, 196
Washers	47	39
Watch chains	340	279
Windmills	257 (b)	194
Wire:		
Strings for musical instruments	59	185
Brushes, rotary	257 (b)	194
Yards	242	192
Steingut (faience)	20 (a), (b)	11
Steinzeug (stoneware)	19 (a)–(d)	11
Stems of flowers in use as drugs	81	57
Stencil ink	85	64 (b)
Sheets		188
Stereotype paper	187	150, 151
Stethoscopes	54 (e)	303
Sticks:		
Golf		297
Joss	105 (b)	87 (b)
Sword	343	44 (c)
For umbrellas and parasols	343	169 (a) 298 (d)
Walking. (According to material.)		
of rattan or bamboo, cut into lengths for manufacturing purposes		169 (a)
Stock, paper, free		315
Stockfish	273	214
Stockholm tar	77 (a)	56 (a)
Stockings, hose, and half hose. (According to material.)		
Stocks and dies	46	194
Stones		Class 1
Artificial	2	2
Common, natural, or artificial	2	2
In the rough, unwrought	2 (a)	2 (a)
Crushed	2 (b)	2 (b)
In millstones, grindstones, whetstones, oilstones, and hones	2 (d)	3
Slabs, plates, or steps	2 (c)	2 (b)
Wrought into all other articles	2 (e)	2 (c)
Fine, such as marble, onyx, jasper, alabaster, and similar fine stones—		
In the rough or squared only	1 (a)	1 (a)
Slab, plates, or steps, polished or not	1 (b)	1 (b)
Sculptures, high and bas-reliefs, vases, urns, and similar articles for house decoration	1 (c)	1 (c)
Wrought into other articles	1 (d)	1 (c)
Oil	1	3
Precious	24	13
Imitations of	26	14
Semi-precious		13
Stoneware, earthenware, faience, porcelain, and ceramic wares		Class I, Group 1
Stools, piano		185
Stops, bench	46	194
Storage batteries	249	193 (a)
Stovepipe	36 (c)	31 (b)
Stoves:		
Of copper	69 (a), (b)	50
Of cast iron	32, 33	29
Electric	248	193 (b)
Of tin plate	57	46 (a), (b)
Of wrought iron or steel	59, 58	47
Straps, shawl, or whatever material	229	288
Straw:		
Caps of	357	301
For forage		228

	1905. Par.	1909. Par.
Straw—Continued.		
Other—		
Manufactured	204	169 (c)
Unmanufactured	203	169 (a)
Hats, bonnets, and crowns of	355	299
Paper	178	150, 151
Sandals	227 (b)	180 (b)
Not otherwise provided for		228
Strawboard		150, 151
Manufactures of		150 (b), 151 (b)
Straws (so-called) of paper	190 (d)	151 (b)
Strings:		
For musical instruments 29 (b), 59, 69 (b),	174, 233	185
For violin bows, of horsehair, prepared	164	185
Hat, manufactured of silk	174	148
Strontium:		
Lactate	99	81
Iodide	99	81
Nitrate	94 (h)	73 (c)
Strophantus seeds	81	57
Structural wrought iron or steel	41	34
Strychnine, and its salts	96	78
Stuccowork of gypsum	4 (a)	7
Stuffed birds and animals	232	289, 290, 344
Stylographic pens	352 (b)	295
Styptic pencils	99	81
Stypticin	96	78
Styrax, a vegetable juice	78	56 (b)
Submarine telegraph cables	395	323
Substances not otherwise provided for		306
Succory root (Chicory)	292	246
Succini oil	99	87 (a)
Sugar:		
Burnt for coloring	85	61 (c), (d)
Export duties on	401	357
Of lead	95 (a)	75
Machinery for making	245	194 (a)
Of milk	99	81
Mills	245	194
Raw	289 (a)	240 (a)
Refined	289 (b)	240 (b)
Sirups		241
Sulphates:		
Of alkaloids	96	78
Of ammonium	94 (d)	74 (a)
Of copper	94 (e)	74 (c)
Of copper, for cauterizing	99	81
Of iron	94 (c)	74 (c)
Of magnesium	94 (c)	74 (c)
Of potassium	94 (e)	74 (a)
Of quinine	384	322
Of sodium	94 (c)	74 (c)
Sulphide:		
Of antimony	94 (h)	74 (c)
Of potassium	94 (h)	74 (c)
Sulphite:		
Of sodium	94 (h)	74 (c)
Of zinc	94 (h)	74 (c)
Sulphocyanide of ammonium	97	77
Sulphonal	99	81
Sulphur	90 (a)	68
Sulphur dioxide		70 (b)
Sulphuret of potash	97	77
Sulphuric:		
Acid	91 (a)	70 (a)
Anhydride	97	77
Ether—		
Commercial	97	77
Anesthetic	99	81

	1905. Par.	1909. Par.
Tan bark	79	66 (a)
Tanigeno	99	81
Tannalbin	99	81
Tannic acid	92 (d)	71 (b)
Tannigen	99	81
Tannicol	99	81
Tanniform	99	81
Tansan mineral water	312 (c)	268
Tansy oil	99	87 (a)
Tape:		
Admixture of materials in		Rule 7, Sec. 5
Cotton	131 (a)	113 (a)
Insulating, electric	248	193 (a)
Linen, hemp, jute, and other vegetable fibers	157	131 (a)
Measures of steel	46	194
Silk	174	148
Wool	166	143
Tapestry. (According to material.)		
Tapioca	284	225
Tar:		
Juniper	99	81
Mineral	7	22
Oil of, pharmaceutically prepared	99	81
Oils, unrefined		22
Stockholm and similar	77 (a)	56 (a)
Tarcolin	97	77
Tarpaulins. (According to material.)		
Tartar:		
Cream of	97	76
Emetic	99	81
Tartaric acid	92 (a)	71 (b)
Tartrate of antimony and potassium	99	81
Tartrates	95 (b)	75
Tassels (trimmings). (According to material.)		
Tea	301	247
Teaberry oil	105 (a)	87 (a)
Teak:		
Timber	192 (a)	157
Wood furniture	196, 197	163, 164
Teeloil	100 (b)	83
Teeth, artificial, with plates or not	342 (b)	282
Telegraph cables, submarine	395	323
Apparatus	248	193 (a)
Other	65	49 (c)
Telephones, and apparatus pertaining to same	248	193 (a)
Telescope cases, receptacles for personal effects	204	288
Telescopes, instruments	365	303
Tenders, for locomotives	246	194
Tendons:		
Aseptic as surgical ligatures	99	82
For medicinal purposes		59
Tennis:		
Balls	352 (d)	297
Nets—		
Of cotton	144	97
Of other vegetable fibers	144	120 (d)
Other. (According to material).		
Rackets		297
Tenoners and tenoning machinery	257 (b)	194
Terebenthinae	77 (b)	63
Terne plate		31
Manufactures of		46
Testing sets, electrical	248	193
Test tubes, glass	16 (b)	20 (e)

	1905. Par.	1909. Par.
Textiles:		
Of cotton		Class V, Group 3
Of metallic threads, exclusively		Rule 10
Of other vegetable fibers		Class VI, Group 2
Of silk		Class VIII, Group 2
Of wool		Class VII, Group 3
Treatment of		Rules 1–11
Surtaxes on—		
For admixtures of materials		Rules 3–11
For bleaching, corresponding provisions of		Class VI
For brocheing		Rule 8
For dyed yarns, corresponding provisions		Classes V, VI
For embroidering		Rule 9
For making up		Rule 11
For metallic threads		Rule 10
For printing, corresponding provisions		Classes V, VI
For stamping, corresponding provisions		Classes V, VI
For trimming		Rule 9
Uneven, ascertainment of thread count		Rule 1, sec. 5
Theatrical costumes and equipment, used, imported with theatrical troupe, temporarily	386	342
Theine	96	78
Theobroma (cacao butter)	290 (b)	248 (b)
Theobromine	96	78
Theodolites (surveying instruments)	365	303
Therapion	99	81
Theriac in powder	99	81
Thermocauteries	54 (e)	193 (b)
Thermometers	365	303
Thermostat, electrical	248	193
Thiocolum	99	81
Thread count, how ascertained		Rule 1, sec. 5
Thread counter, use of		Rule 1, sec. 5
Thread for sewing:		
Cotton	116	95
Linen and other vegetable fibers	144	120
Silk	170, 170 (a), 171 (c), 172	146
Thread, silver	28 (d)	26 (d)
Three-in-one oil	97	77
Threshing machines	245	194
Thyme:		
Ground	296 (b)	255 (b)
Unground	81	57
Oil of	105 (a)	87 (a)
Thymol	99	81
Tickets, lottery, advertisement of, and list of drawings in, prohibited		Sec. 6 (f)
Tiger-eye (stone)		13
Tiles, ceramic	18 (a), (b)	10
Timber	192	157, 158
Timsim (vegetable pith)	366	169 (a)
Tin and alloys thereof:		
Capsules for bottles	72 (c)	53 (b)
In, bars, sheets, pipes, and wire	72 (b)	53 (a)
In foil	72 (c)	53 (a)
In ingots or lumps, pure	72 (a)	309
Alloys		53 (a)
Solder, in bars	72 (b)	53 (a)
Trinkets of	340	279
In other articles	72 (d)	53 (b)
Tin, muriate of	94 (h)	74 (c)
Tin plate or terne plate, in sheets	36 (b)	31 (c)
Manufactures of	57	46
Tires:		
For bicycles and velocipedes	252	199
For automobiles	352 (d)	198

	1905. Par.	**1909.** Par.
Tramway:		
Cars for freight	263	200 (a)
Other		200 (b)
Transformers, electrical	250	193 (a)
Choke coils	250	193 (a)
Transit, merchandise in, how treated when act becomes		
effective		Sec. 3
Transits		303
Traps, lead, sanitary		54 (a)
Trass	3 (b)	6 (b), (c)
Traveling bags, of whatever material	228 (c)	288
Traveling rugs, making up defined, surtax for		Rule 11, sec. 5
Travelers, personal effects of, when free		339, 340
Trays, developing (photographic)		302
Trebenum	99	81
Trees:		
Christmas, artificial		296
Live	367	308
Trimmings. (According to material.)		
Admixtures of materials in		Rule 7, sec. 5
Or embroidery, applied to textiles, surtax for		Rule 9, sec. 5
Trinkets and ornaments, except those of gold, silver,		
platinum, gold or silver plate, or of amber, coral, ivory,		
jet, meerschaum, mother-of-pearl, and tortoise shell	341 (b)	ˉ279
Triocol	99	81
Trional	99	81
Tripoli (earth)	3 (b)	6 (b), (c)
Tripods for cameras	362	302
Trophies, when free		330
Trowels	46	194
Trucks, handcarts, and wheelbarrows	265 (a)	197
Trunks, of whatever material		288
Tube expanders	46	194
Tubes and pipes:		
Of clay	17 (d), (c)	9
Glass, test	16 (b)	20 (e)
For insulation		193 (a)
Other. (According to material.)		
Tubs. (According to material.)		
Tulles. (According to material.) (*See* Rule 6.)		
Admixtures of materials in		Rule 6, sec. 5
Tumblers, glass	13 (a), (b)	16
Tungtree oil	100 (b)	83
Tuns, pipes, casks, and barrels	363	161
Turbans	356	301
Turbine engines	243	194
Turkeys	268	175, 204
Canned, or potted	315, 316	210, 211
Turkish towels and bath robes:		
Of cotton pile fabrics	124 (a)	106
Other. (According to material.)		
Turntables	247	194
Turpentine, oil or spirits of	77 (b)	63
Turpentine, Venetian	78	56 (b)
Twine:		
Of cotton, for wrapping and for sewing sails	116 (a)	96
Of hemp and other vegetable fibers	144, 146	120
Other. (According to material.)		
Twisted silk	170	146
Type:		
Of lead	73 (f)	54 (a)
Metal. (*See* Lead.)		
Typewriters and parts, including ribbons	253	188
Typewriting paper	179	150, 151
Typewritten documents, free		329

	1905. Par.	1909. Par.
Ultramarine blue:		
Dry	84 (c)	61 (c)
In liquid or paste		61 (d)
Umber:		
Dry	83	60
In liquid or paste	84 (d)	61 (d)
Umbrellas and parasols:		
Covered with paper	354 (a)	298 (a)
Covered with silk	354 (b)	298 (b)
Covered with other stuffs	354 (c)	298 (c)
Sticks for	343	169 (a), 298 (d)
Frames for	59	47, 298 (d)
Handles. (According to material.)		
Underwear, knit		Rule 6
Of cotton	125	107
Of linen and other vegetable fibers	152	127
Of wool	165	141
Of silk	173	148
Underwriters, when recognized as consignees		Sec. 17
United States:		
Government supplies	385	338
Unusual coverings, containers, or packing		Rule 13, (h), (i), Sec. 5
Unwashed wool	162 (a)	314
Uranium, acetate, chloride, nitrate, oxide	99	81
Urns:		
Decorative, of bisque, clay, faience, porcelain, stoneware	23	11 (f)
Of glass	13 (a), (b)	20 (b), (c)
Of marble, onyx, jasper, alabaster, and similar fine stones	1 (c)	1 (c)
Other. (According to material.)		
Urotropin	99	81
Usual coverings, containers, or packing		Rule 13, Sec. 5 (f), (g), (h)
Vaccines and serums	99	324
Vacuometers		303
Vacuum automatic brakes	257 (b)	194
Valerianate of zinc	99	81
Valerian oil	99	87 (a)
Validol	99	81
Valises of whatever material	228 (c)	288
Value, dutiable, definition of		Rule 13 (a), Sec. 5
Market, defined		Rule 13 (a), (b), Sec. 5
Valves:		
Essential parts of machines		194
Other (not machinery). (According to material.)		
Vaniline, or vanillin	97	77
Vanilla:		
Beans	295	259
Flavoring extracts	327	258
Varnishes	88	62
Varnished hides, skins, and leathers	218	178 (c)
Vaseline:		
Plain	10	23 (b), (c)
Preparations of, for toilet purposes	105 (b)	87 (b)
Preparations of, for medicinal purposes	99	81
Vases. (According to material.)		
Vasogena	99	81
Vegetable:		
Fibers, cotton		Class V
Other		Class VI
Flavoring extracts	327	258
Fuel	201	167
Gums and resins		56

	1905. Par.	1909. Par.
Vinegar	314	257
Vinolia powder and cream	105 (b)	87 (b)
Violet oil	105 (a)	87 (a)
Violin strings	59, 174, 233	185
Violins	195, 196, 197	185
Viridis soap	104	86
Vises	46	194
Visiting cards	179, 190 (a)	150, 151
Vitriol	91 (a)	70 (a)
Volatile or essential oil. (*See* Oils.)		
Voltmeters	248	193 (a)
Vulcanized oil	99	81
Wads for shotguns:		
Cardboard	190 (a)	150, 151
Felt	166	139, 143
Wafers:		
Unsweetened	283 (a)	223 (a)
Sweetened	283 (b)	223 (b)
Wagons and carts for transporting merchandise	264	197
Other	259	201
Wagon wheels of wood	266	202
Walking sticks. (According to material.)		
Wall paper	184	151
Wall telephones and parts thereof	248	193 (a)
Walnut oil	100 (b)	83
Walnuts		239
Ware, hollow:		
Ceramic	19–21	11
Enameled wrought iron	59	47 (b)
Wedgewood	19	11
Other. (According to material.)		
Warehouse, and hand trucks, wheelbarrows	265 (a)	197
Warp, definition of		Rule 1, sec. 5
Washed wool	162 (b)	314
Washers:		
Copper	68 (a), (b)	50
Rubber	352 (a)	293 (a)
Wrought iron or steel	47	39
Other. (According to material.)		
Washing machines	257	194
Washing soda	94 (e)	74 (c)
Waste:		
Copper, iron, steel, and other common metals	74	309
Cotton	112	93
Of other vegetable fibers	365	305
Silk	168	144
Wool	162 (c)	137 (a)
Wastes:		
Animal, and by-products	234	184
Common metals, fit only for resmelting	74	309
Not otherwise provided for		305
Watches, clocks, parts of and accessories for	238	187
Watchmen's clocks	239	187
Water:		
Aerated, artificial, or natural mineral	312 (c)	268
Ammonia	93	73
Caltrops, nuts	329	239
Colors	85	61 (c), (d)
Meters	257 (a), (b)	194
Mineral, natural, aerated, carbonated, or not	312 (c)	268
Perfumed toilet	105 (b)	87 (b)
Sweetened, flavored, or aerated	312 (c)	268
Waterproof and caoutchouc stuffs:		
On cotton textiles	135	117
Linen and other vegetable fibers	160	135

	1905. Par.	1909. Par.
Windlasses, steam	245	194
Windmills, steel	257 (b)	194
Window glass, common	14 (b), (c)	18 (b)
Plate		18 (c)
Wines:		
Chinese	308, 310, 311	265, 266, 263
Heavy oil of	105 (a)	87 (a)
Lees		76
Medicated	99	80, 81
Sparkling	309	264
Vermouth	310	265, 266
Wines, still	310, 311	265, 266
Wintergreen, oil of, natural and artificial	105 (a)	87 (a)
Wire:		
Copper, brass, bronze, etc., blanched, gilt, or nickeled		49 (b)
Cables for electricity	65	49 (c)
Covered with textiles other than silk	65	49 (c)
Covered with silk	65 (a)	49 (d)
Gauze	66 (a), (b)	49 (e)
Manufactures of		49 (f)
Plain		49 (a)
Strings for musical instruments	69 (a), (b)	185
Gold	27 (d)	25 (d)
Nickel, aluminum, and their alloys	71 (b)	52 (a)
Platinum	27 (d)	25 (d)
Silver	28 (d)	26 (d)
Tin and alloys thereof	72 (b)	53 (a)
Wrought iron or steel—		
Barbed		36 (a)
Cables and ropes		36 (a)
Covered with textiles	40 (c)	36 (c)
Galvanized	40 (a), (b), (c)	36 (a), (b)
Gauze, cloth or screenings	44 (a), (b)	36 (d)
Manufactures of		36 (e)
Nails	48	40
Netting	45	36 (b)
Nickeled or bronzed		36 (c)
Plain		36 (a), (b)
String for musical instruments	59	185
Zinc, lead, and other metals	73 (b)	54 (a), Class X
Wood:		
Alcohol	307	260
In barbers' and dentists' chairs	199	165
In bar fixtures	199	162, 163, 164
Beads of, in necklaces or other trinkets	340	279
Loose	195, 196, 197	162 to 164
Bent, furniture of	198	162
In billiard, pool, or bagatelle tables	200	166
In buttons	345 (a)	283 (b)
Carved	197	164
Cedar	192 (a)	157
Common (acacia, alder, ash, Australian jarra, beech, birch, black poplar, California redwood, cedar, cypress, elder, evergreen oak, maple, oak, pear, pine, plantain, poplar, spruce, and yew-leaved fig)	192	157
Unmanufactured, shingles, laths, and fencing		157
Manufactured		162, 163, 164
Coopers' wares, in shooks, staves, coops, headings, and bungs	194	160
Dominos of, toys	353	296
Excelsior		159
Fencing		157 (b)
For dyeing	86 (a)	66 (a)
Fillers		62

	1905. Par.	1909. Par.
Yarn:		
Of cotton		94, 95
Of other vegetable fibers		119, 120
Of silk		146
Of wool		138
Yeast	97	77
Yellow ocher	83	60
Yohimbina	96	78
Zimine	99	81
Zinc, and alloys thereof:		
Articles which are bronze gilt or nickeled	73 (c), (d)	54 (c)
Bars, bar solder, pipes, sheets, traps, and type	73 (b)	54 (a)
Ingots or lumps, pure	73 (a)	309
Alloys		54 (a)
In plain articles	73 (f)	54 (b)
Wire	73 (b)	54 (a)
Zinc:		
Carbonate	94 (h)	74 (c)
Chloride	94 (h)	74 (c)
Iodide	99	81
Oxide	84 (c), (d)	61 (c), (d)
Phosphate	99	81
Sulphate	94 (h)	74 (c)
Valerianate	99	81
Zincs for batteries, dry or wet	248	193 (a)
Zoology specimens for public museums, schools, academies, and scientific corporations	389	344

O

CPSIA information can be obtained
at www.ICGtesting.com
Printed in the USA
BVHW04*1142031018
529148BV00014BA/293/P